VIROLOGY MONOGRAPHS

DIE VIRUSFORSCHUNG IN EINZELDARSTELLUNGEN

CONTINUING / FORTFÜHRUNG VON

HANDBOOK OF VIRUS RESEARCH

HANDBUCH DER VIRUSFORSCHUNG

FOUNDED BY / BEGRÜNDET VON

R. DOERR

EDITED BY / HERAUSGEGEBEN VON

S. GARD · C. HALLAUER · K. F. MEYER

3

1968

Springer-Verlag Wien GmbH

CYTOMEGALOVIRUSES

BY

J. B. HANSHAW

RINDERPEST VIRUS

BY

W. PLOWRIGHT

LUMPY SKIN DISEASE VIRUS

BY

K. E. WEISS

1968

Springer-Verlag Wien GmbH

ISBN 978-3-662-38850-1   ISBN 978-3-662-39771-8 (eBook)
DOI 10.1007/978-3-662-39771-8

© 1968 by Springer-Verlag Wien
Ursprünglich erschienen bei Springer Vienna 1968.
Softcover reprint of the hardcover 1st edition 1968

Library of Congress Catalog Card Number 68-26922

Printer: Steyrermühl, A-1061 Wien, Austria
Title No. 8330

Cytomegaloviruses

By

J. B. Hanshaw

University of Rochester School of Medicine and Dentistry
260 Crittenden Boulevard, Rochester, New York 14620, U.S.A.

With 6 Figures

Table of Contents

I. Introduction and History

Since the isolation of the first human strains of the cytomegalovirus (CMV) more than a decade ago, it has become increasingly apparent that these and related agents are common infections for man and other animal species. Infection in man may or may not induce an illness — cytomegalic inclusion disease (CID). The latter terminology is derived from the observation that infection *in vitro* and *in vivo* results in marked cytomegaly. This is due in part to the presence of an usually large, nuclear inclusion. An alternate term, salivary gland virus (SGV), has been used in the earlier literature to draw attention to the fact that the distinctive inclusions were frequently found in the lining epithelial cells of the salivary ducts. WELLER and HANSHAW (1962), MEDEARIS (1964), HANSHAW (1966a), have established that the human CMV are significant fetal pathogens capable of inducing oculo-cerebral defects as well as a variety of extraneural abnormalities. In addition, the infection in man and in other species has been characterized by latency, chronic infection and reactivation not unlike that observed in the closely related *herpesvirus hominis* (herpes simplex) and *herpesvirus varicellae* (varicella-zoster). ANDREWES (1962) and LWOFF, HORNE and TOURNIER (1962) have suggested that the cytomegaloviruses be included in the herpesvirus group. In view of the close biological, physical and chemical similarity of these agents their ultimate classification as one group appears probable.

The characteristic cytomegalic cells were first noted by JESIONEK and KIOLEMENOGLOU (1904) in the kidneys, liver and lungs of a newborn infant regarded as syphilitic. Similar cells were noted concurrently by RIBBERT (1904) in the kidneys of a newborn infant and in the parotid gland of two older infants studied at autopsy. LÖWENSTEIN (1907) reported cytomegaly in the parotid ducts of 4 of 30 infants dying of a variety of illnesses. The inclusions were regarded by some authors as protozoa; others considered the changes syphilitic in origin. GOODPASTURE and TALBOT (1921) suggested the possibility of chronic inflammation and noted the similarity of the observed "giant cells" to the cytomegaly described in the salivary glands of guinea pigs. Similar, but smaller inclusions were observed by TYZZER (1906) in biopsies of varicella skin lesions. LIPSCHÜTZ (1921) recognized the similarity of the intranuclear inclusions to those seen in herpetic lesions and postulated a viral etiology for the disease.

The first experimental evidence to support the viral etiology of the infection resulted from work in guinea pigs by COLE and KUTTNER (1926). They were able to induce inclusion body formation in young animals injected with salivary gland material obtained from older guinea pigs. The inoculum remained infective after passage through a Berkefeld N filter but was not infective after heating. Material from the guinea pig did not induce cytomegaly in other species and other rodents with cytomegaly could not transfer the infection to a different species.

FARBER and WOLBACH (1932) demonstrated cytomegaly restricted to the salivary glands of 12% of 183 infants dying of all causes. The presence of this localized cytomegaly was not limited to one clinical syndrome but rather seemed to appear irrespective of the cause of death. During the last two decades an increasingly large number of cases of generalized CID have been recognised. This is largely due to the development of tissue culture methods for the isolation of the virus from living patients.

SMITH (1954) succeeded in propagating the murine cytomegalovirus in explant cultures of mouse embryonic fibroblasts. The isolation of the human cytomegalovirus was accomplished shortly thereafter by SMITH (1956), ROWE et al. (1956), and WELLER et al. (1957). SMITH isolated the agent from two infants, one dying of generalized cytomegalic inclusion disease. ROWE and his associates isolated 3 strains of virus from degenerating adenoid tissue of children undergoing adenoidectomy. WELLER et al. recovered virus from the urine and liver of living infants with generalized cytomegalic inclusion disease. In each laboratory an identical cytopathic effect was demonstrated. Viral replication was restricted to human fibroblastic tissue and was characterized by the development of large, refractile cells occurring focally. On staining, these cells contained intranuclear inclusions similar to those observed earlier in patients dying with localized and generalized infection. With the availability of an *in vitro* method for the propagation of the virus, serologic methods, including the neutralization and complement fixation (CF) tests, were developed. Antigenic heterogeneity noted in cross-neutralization tests by WELLER, HANSHAW and SCOTT (1960) have been less apparent in CF tests. Employing a microtechnique SEVER (1963) and MEDEARIS (1964) have demonstrated that the complement fixation test can be useful in diagnostic and epidemiologic investigations. Evidence has been presented by HANSHAW et al. (1965) to suggest that cytomegalovirus infection acquired after birth may produce liver abnormalities in normal individuals as well as those with impaired immune mechanisms.

Recent reviews include those by MEDEARIS (1964), WELLER (1965), and HANSHAW (1966 b).

II. Classification and Nomenclature

With the evolution of a system of viral classifications such as that proposed by ANDREWES (1962) and LWOFF, HORNE and TOURNIER (1962) it has become clear that the cytomegaloviruses are closely related to the herpesviruses. The system of classification adapted in 1965 by the Provisional Committee for Nomenclature of Viruses is based on:

1. nucleic acid type;

2. symmetry;

3. presence of an envelope;

4. the triangulation and capsomere numbers.

The cytomegaloviruses are identical to the known herpesviruses in each of these characteristics, and in this classification are listed as a separate genus (cytomegalovirus) within the *herpesviridae* family. Included in this genus are species specific non-human cytomegaloviruses which have been isolated from the mouse by SMITH (1954) and the guinea pig by HARTLEY et al. (1957). In addition, a large number of animals including rats, sheep, moles, hamsters, chimpanzees, dogs and pigs have been found to have cytomegaly probably due to related agents.

WELLER, HANSHAW and SCOTT (1960) proposed the term "cytomegalovirus" for the human SGV agents because:

1. the salivary glands are only one of many possible sites of involvement; and

2. unrelated agents obtained from bats have also been designated as salivary gland viruses.

On the basis of cross-neutralization tests WELLER, HANSHAW and SCOTT (1960) suggested that the cytomegaloviruses be divided into two serotypes represented by the DAVIS (type I) isolated by WELLER et al. (1957) and the AD 169 (type II) strains isolated by ROWE et al. (1956). A third possible serotype represented by the Esp. and Kerr strains (WELLER et al., 1957) was found to be closely related but showed some cross-reactivity with the DAVIS strain. The type II prototype (AD 169) originally isolated by ROWE et al. (1956) has been widely used as the antigen source in complement fixation tests.

III. Properties of the Virus

A. Morphology

The most characteristic morphologic feature of cytomegalovirus infection is the large intranuclear inclusion (Fig. 1). The inclusion-bearing cell has an over-all diameter of 25—40 microns with an inclusion diameter of approximately 8—10 microns (SMITH, 1959). Both the nucleus and the cytoplasm contribute to the striking increase in the size of the infected cell. The nuclear inclusion is composed of chromatin aggregates and three types of virus particles. It is located centrally and may correspond in shape to that of the cell. With hematoxylin and eosin staining the inclusion may be eosinophilic or basophilic but tends to be more eosinophilic than the prominent nuclear membrane. The latter structure is thickened and usually has a beaded appearance thought to represent aggregations of nuclear chromatin. The central inclusion is surrounded by an electron-lucent halo which is clearly seen in preparations examined with the light microscope. Unlike the uninfected fibroblast which has a centrally placed nucleolus, the CMV infected cell has this structure displaced to the nuclear periphery. The nucleolus of the infected cell does not contain viral particles. Electron micrographs of a single CMV particle and a group of particles in the cytoplasm are shown in Figs. 2 and 3.

PATRIZI et al. (1965) describe three virus particles in the nucleus. One is a spherical particle 65 to 85 mμ in diameter with a limiting membrane. A second virus form is spherical and consists of an electron-lucent or electron-opaque core surrounded by an outer (60—80 mμ in density) and inner (35—40 mμ in diameter) membrane. Both virus particles are associated with aggregates of chromatin in the nuclear inclusion. A third viral particle has an additional outer coat with an outside diameter of 115—145 mμ. The core is electron-opaque and is surrounded by an electron-lucent area (60 to 75 mµ in diameter). These particles are frequently found in clusters near the nuclear periphery and may be seen between the inner and outer nuclear membranes.

Cytoplasmic inclusions, 2—4 microns in diameter, are sometimes seen in cells containing nuclear inclusions but have not been described in the absence of nuclear changes. These inclusions are usually basophilic and react positively with periodic acid-leukofuchsin (PAS stain). Recent studies by McGAVRAN and SMITH (1965), RUEBNER et al. (1966), PATRIZI et al. (1965), have found that these inclusions contain mature virus particles surrounded by larger non-viral particles which

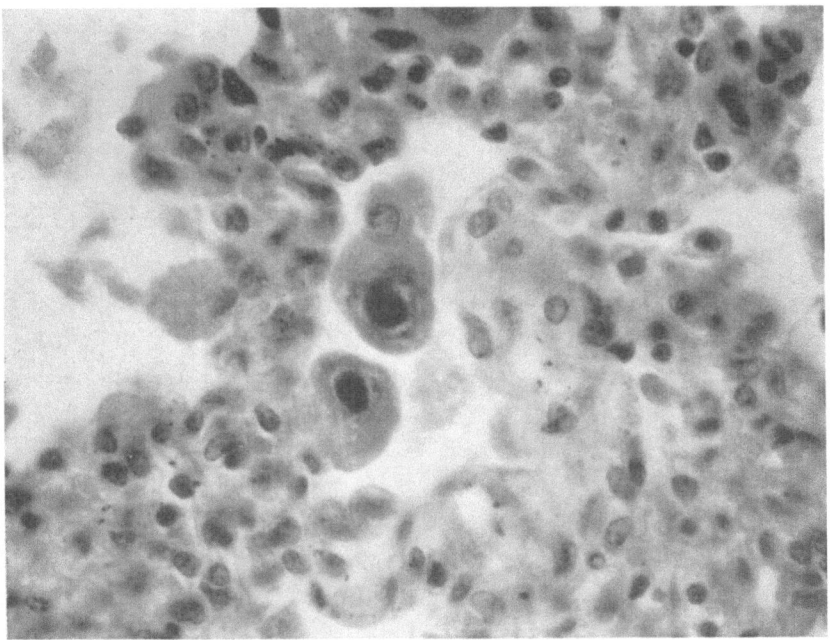

Fig. 1. Two large cells with intranuclear inclusions in a child with cytomegalovirus pneumonitis. Reproduced from HANSHAW (1964) with the permission of the publisher.

are thought to be lysosomes. HORN et al. (1964) have postulated that the lysosomes engulf and surround the virus particles in a manner comparable to the phagocytosis of bacteria. This may reduce the infectivity of the virus and possibly account for the low infectivities that occur *in vitro*. RUEBNER et al. (1966) have suggested that the relatively low lysosomal content of the salivary glands may account for the persistence of murine cytomegalovirus in this organ following intraperitoneal injection of virus.

The mature virus particles seen in the cytoplasmic inclusions are similar to the nuclear particles located between the inner and outer nuclear membranes. The outer coat is somewhat thicker in the cytoplasmic than the nuclear form. BECKER et al. (1965) have measured average diameters of nuclear and cytoplasmic particles and found that nuclear particles averaged 111 mµ in diameter in contrast to 142 mµ for cytoplasmic particles.

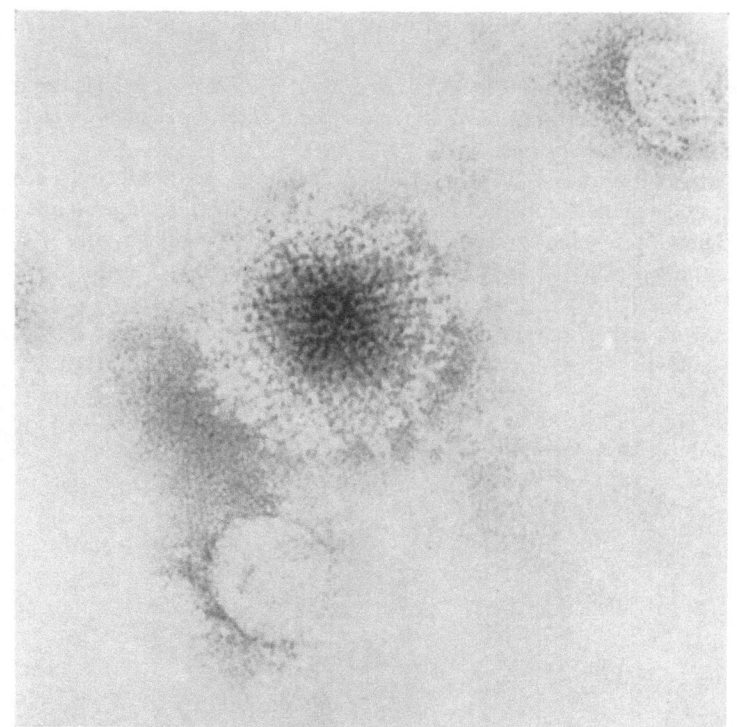

Fig. 2. Electronmicrograph of a single CMV particle (504,000 ×). Reproduced with the permission of Drs. H. R. MORGAN and P. BALDUZZI.

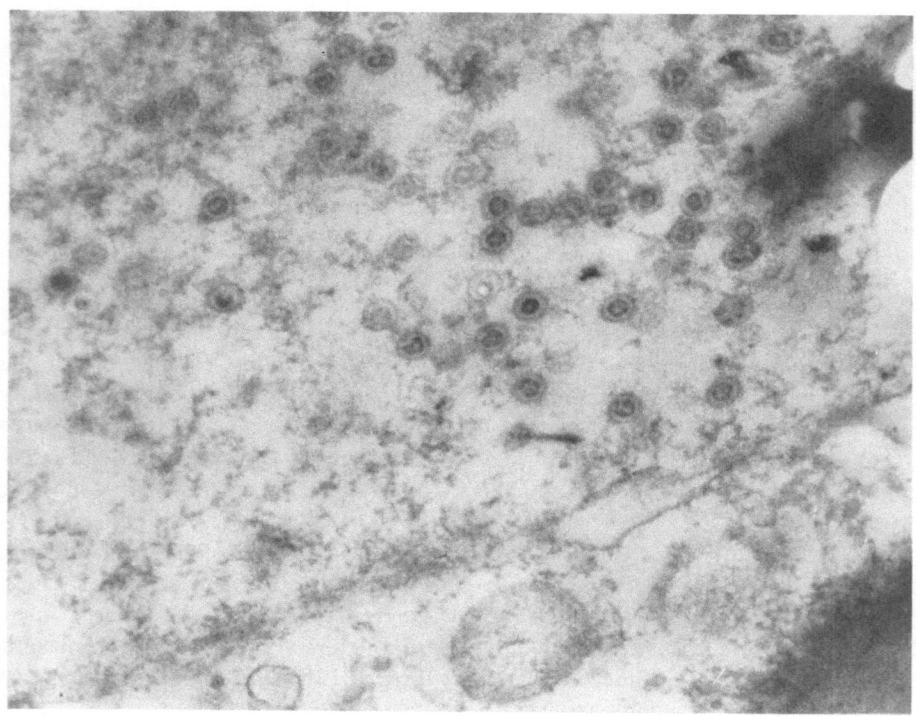

Fig. 3. Electronmicrograph of a group of cytoplasmic CMV particles (100,000 ×). Reproduced with the permission of Drs. H. R. MORGAN and P. BALDUZZI.

Several investigators have attempted to explain the low infectivity of CMV in terms of electron microscopic observations. SMITH and RASMUSSEN (1963) have noted that CMV has a high particle-infectivity ratio, approximating 10^7 to 10^8. Although CMV infected cultures had an average yield of 10,000 particles per cell, they found only one infectious unit per 1000 cells. They suggest that this inefficiency of viral synthesis is explained, at least morphologically, by the incompleteness of the nucleic acid cores. Not one particle in over a thousand stained as if it had a complete core. BECKER et al. (1965) have observed the same phenomenon in electron micrographs of varicella-zoster virus and observe that both CMV and V-Z virus continue to apply protective envelopes to particles which apparently lack sufficient nucleic acid.

The basic morphology of the cytomegaloviruses is well documented (LUSE and SMITH, 1958; STERN and FRIEDMAN, 1960; SMITH and RASMUSSEN, 1963; RUEBNER et al., 1964; 1966; BECKER et al., 1965; McGAVRAN and SMITH, 1965; PATRIZI et al., 1965). BECKER et al. (1965) have not noted marked differences in the intracellular development and particle morphology among the CMV and members of the herpesvirus group. The only significant difference noted by SMITH and RASMUSSEN (1963) is the unusually large number of completely or partially empty nucleoids in CMV particles as compared to HSV. Both HSV and CMV had 162 capsomeres (22—23 around the periphery), and the average distance between capsomeres was 125 Å from center to center. BECKER et al. (1965) noted that V-Z virus acquired envelopes from already existing membrane systems, namely, the endoplasmic reticulum, while CMV particles acquired their protective envelope from the limiting membrane of the cytoplasmic inclusion. Both viruses also acquired coats at the nuclear membrane. PATRIZI et al. (1965) believe that this coat is derived from the inner nuclear membrane and that the outer nuclear membrane evaginates into the cytoplasm. It is uncertain whether the outer nuclear membrane contributes to the virus coat.

B. Physico-chemical Structure

Detailed knowledge of the chemical structure of CMV is not available. On the basis of staining with phosphotungstate, uranyl acetate (SMITH and RASMUSSEN, 1963), acridine orange (NIVEN, 1959), and the inhibitory effect of the DNA antagonist, 5-fluorodeoxyuridine (GOODHEART et al., 1963), the DNA nature of the core is well established. The molecular weight of the DNA has been estimated by GREEN (1965) to be 64 million with a guanine and cytosine content of 58% (CRAWFORD and LEE, 1964). Herpes simplex, pseudorabies, and infectious bovine rhinotracheitis viruses contain 68 to 74% guanine and cytosine while equine abortion, LK, and CMV are in the 56 to 58% range. GREEN postulates that these differences may form the basis for natural subgroups.

The virus particle contains in addition to the core DNA, an icosahedral capsid (1000 Å in diameter) composed of hollow, elongated capsomeres made up of 960 structural units, 12 pentomers and 150 hexamers. Essential lipids are present in the virion. Perhaps observations such as those made by SMITH (1963) on herpes simplex virus can be applied to the cytomegaloviruses. The HSV envelope is stripped from the capsid upon treatment with ether, exhibiting no visible damage to the capsid. The naked capsid is not infective without the envelope.

WATSON amd WILDY (1963) found that only the complete HSV virion with envelope is agglutinated by anti-host cell serum indicating that the envelope
consists of host cell material.

C. Antigenic Structure

The CMV have neutralizing and complement fixing antigens (WELLER et al.,
1957; ROWE et al., 1956). Although a given CF antigen may be broadly reactive
against antibody produced by many strains, there is less antigenic homogeneity
in neutralization tests. As noted above, WELLER, HANSHAW and SCOTT (1960)
have demonstrated two, and possibly three, serotypes. HANSHAW et al. (1965)
found that type I neutralizing antibody (prototype Davis) was more widely
distributed among 25 CMV infected children than the type II (AD 169) antibody.
The AD 169 strain, however, is a broadly reactive CF antigen (ROWE et al., 1956;
SEVER et al., 1963). The observation that AD 169 CF antibody is present in a
large percentage (81%) of people over 35 attests to its usefulness as a single
antigen in the detection of CMV antibody. In contrast to the cumbersome neutralization test, the CF test has been technically more suitable for sero-epidemio-
logic investigations of large numbers of individuals.

BENYESH-MELNICK et al. (1966) found that the CF antigen of CMV could
be separated from the virus particles by differential centrifugation. The separated
antigen was found to be thermostable at 4°C but unstable at 37°C. This is the
reverse of the infectivity of this virus and is of importance in the performance
of the CF test. It is not known if the soluble CF antigen separated from the virus
particle is a structural protein of the virus which was released by disintegration
of the virus particle or a different protein coded for by the virus genome. BENYESH-
MELNICK et al. suggest that the relatively large amount of CF antigen which
remains in the pellet of the centrifuged virus (50%) favors the first explanation. CMV CF antigen is usually undetectable in the fluid phase preparations
of CMV.

Hemagglutinins have not been demonstrated in CMV infected cells. No
hemolysin is produced and the hemadsorption phenomenon has not been described.

CMV strains isolated from mice and rodents are antigenically distinct from
each other and from human strains in CF and neutralization tests. BLACK et al.
(1963), however, isolated two strains of CMV in human tissue from *Cercopithecus*
monkeys which shared CF antigens with a human CMV (AD 169) strain. Reference antisera to human CMV strains have been prepared in monkeys by repeated
intraperitoneal and intramuscular inoculations of infected trypsinized, human
lung fibroblasts (PLUMMER and BENYESH-MELNICK, 1964).

D. Resistance to Physical and Chemical Agents
1. Fat Solvents

CMV is sensitive to ether (ROWE et al., 1956) and chloroform (HAMPARIAN
et al., 1963). These observations suggest that the virus particle contains lipid
although a chemical analysis of purified virus has not been done. MCALLISTER
et al. (1963) have noted that the cytoplasmic inclusion but not the nuclear inclusion is Sudan IV-positive 24 hours after infection.

2. Heat

The sensitivity of CMV to temperature change has been noted since these agents were first isolated. Recently VONKA and BENYESH-MELNICK (1966) studied the thermoinactivation of CMV under varying conditions. They found that:

1. freshly harvested CMV was more rapidly inactivated at 4°C than at higher temperatures;

2. at 4°C and 37°C cell-associated virus was not more stable than extracellular virus released by sonic disruption of cells;

3. the deletion of bicarbonate from Eagle's medium tended to stabilize CMV at 37°C and 4°C;

4. distilled water has a stabilizing effect at 4°C and 37°C;

5. virus harvested soon after infection was less labile at 37°C than virus harvested later. Of the various strains tested by these authors including AD 169, Davis, and Kerr, AD 169 was the most stable at 4°C. This property may be a factor in the usefulness of this strain as a complement-fixing antigen. No differences in stability were noted at 37°C.

PLUMMER and LEWIS (1965) have found that, unlike herpes simplex, CMV is less stable at 4°C and 10°C than at 22°C. VONKA and BENYESH-MELNICK (1966) have established that CMV penetration of human embryonic lung fibroblasts is temperature-dependent occurring at 37°C but not 4°C. Virus attachment *per se*, however, does not appear to depend on temperature. Prolonged incubation of virus at 4°C will inactivate a high proportion of attached virus and prevent penertation. Cell-associated complement-fixing activity has been shown to be thermolabile at 37°C. Quick freezing in alcohol and dry ice prior to storage at −70°C is recommended. Stability may be enhanced by sorbitol in a final concentration of 25 to 35% (WELLER and HANSHAW, 1962).

3. Acid

The cytomegaloviruses are labile in the presence of solutions below pH 3.0. VONKA and BENYESH-MELNICK (1966), however, have noted increased plating efficiency of CMV when the bicarbonate concentration of the methyl cellulose overlay is reduced from 0.225 to 0.15%. The plaques formed under the more acid overlay were bigger and appeared earlier. Similar results can be achieved in the atmosphere of 10% CO_2 suggesting that the bicarbonate effect is due to lowering of the pH. These authors were not able to explain the increased plating efficiency on the basis of enhanced virus attachment in the acid medium.

E. Cultivation

1. Host-cell Range

With one exception noted above (BLACK et al., 1963), the cytomegaloviruses have been successfully propagated only in the fibroblastic tissue of the species from which the agent was isolated. The fibroblastic preference *in vitro* is in contrast to the tendency for epithelial cell involvement *in vivo*. The tissue culture systems that have been most widely used in the isolation of the human cytomegalovirus have been derived from human embryos. The explant cultures used by

WELLER and his associates (1957) were derived primarily from embryonic skin and muscle but satisfactory cultures have also been derived from myometrium, testes, kidney, and lung. More recently, diploid cell lines obtained from human embryonic lung (WI-38) have been found to be convenient and sufficiently durable to meet the requirement of long term cultivation (HANSHAW, 1966).

Human strains have not been shown to produce disease in non-human hosts (ROWE et al., 1956; WELLER et al., 1957; MEDEARIS, 1964). The species specificity of the cytomegaloviruses is a characteristic shared by varicella zoster but one that serves to distinguish CMV from herpes simplex.

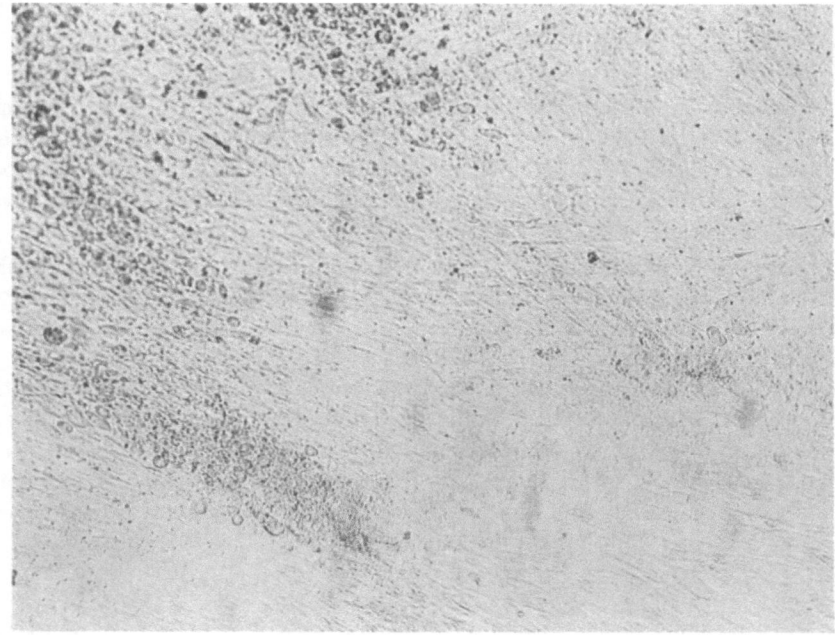

Fig. 4. Cytomegalovirus infection in human embryonic fibroblasts 72 days after inoculation. Four focal areas of rounded cells can be seen. Reproduced from HANSHAW (1964) with permission of the publisher.

2. Virus Multiplication

In contrast to most viruses that can be successfully propagated *in vitro*, CMV replicates slowly. Usually infectious virus cannot be detected in the cell-free media. McALLISTER et al. (1963) were unable to detect virus in either the cells or medium 48 hours after infection. The intracellular virus then accumulates in the cells and is released in the medium. The relatively slow rise in titer of extracellular virus has not been completely explained but could be related to the inactivation of virus at 37°C, the slow release of virus from the cell, or adsorption of released virus to uninfected cells. BECKER et al. (1965) have detected virus particles in the nucleus, cytoplasm, and extracellular fluid 5 days after infection. VONKA and BENYESH-MELNICK (1966) found that cell-associated virus precedes the appearance of virus in the fluid phase by 2—5 days. The development of cytopathic changes follows closely the appearance of infectious virus.

3. Cytopathogenicity

There is a wide range of time (24 hours to 6 weeks) in which the cytopathic effect of inoculated cultures may become manifest. This variable prepatent period is largely a function of the virus titer of the inoculum at the moment of inoculation. A well adapted strain of virus (such as AD 169) will frequently show generalized rounding and refractility within 24 hours. Specimens of urine obtained from infants with evidence of systemic infection (petechiae, jaundice, hepatosplenomegaly) may contain large amounts of infective virus and may also induce cytopathic changes within a 24 hour period. Freshly isolated strains, however, do not usually produce such an immediate effect and characteristically the first

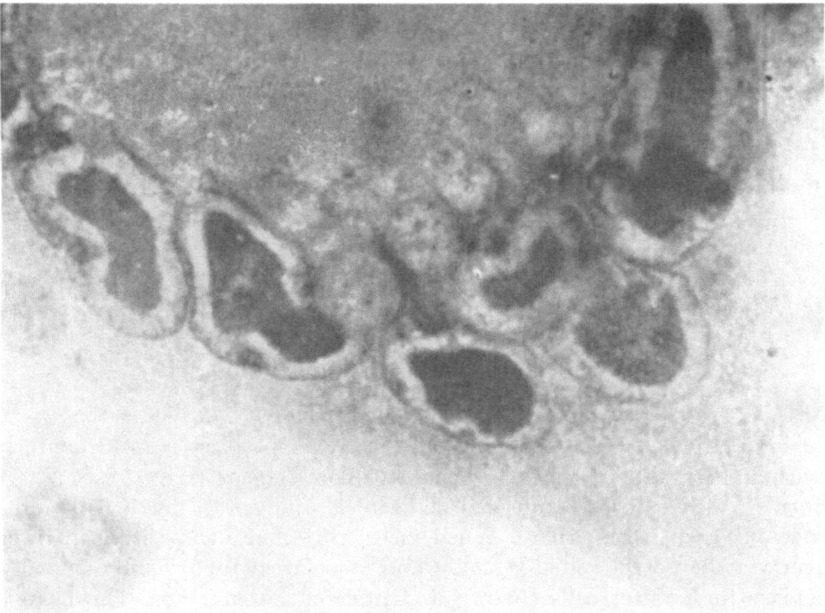

Fig. 5. Cytomegalovirus infection in human embryonic fibroblasts which show intranuclear inclusion with prominent margination of nuclear chromatin.

changes are noted between the 5th and 21st day following inoculation. In such cultures the initial cytopathic changes tend to be focal collections of rounded, refractile cells (Fig. 4). They may vary in number from 2 or 3 in the earliest stages to over 100 foci as the process spreads to involve contiguous cells. Infection continues until the entire cell sheet becomes involved. This is the optimum time for the harvesting of infectious virus, as well as the cell-associated complement-fixing antigen. The appearance of the CMV focal lesion is sufficiently distinct to be considered pathognomonic (WELLER and ROWE, 1964). The greenish-brown, refractile, cytoplasmic granules described by WELLER et al. (1957) in plasma clot cultures are less prominent or absent in monolayer cultures. Similar slowly progressive, focal changes occur in varicella-zoster infected cultures. These focal lesions are somewhat more delicate in appearance and, unlike CMV, the virus can be propagated in epithelial cells of human and simian origin (WELLER and

ROWE, 1964). In contrast, herpes simplex virus progresses rapidly in culture, produces very transient focal lesions, and can be propagated in a wide variety of tissue culture systems and hosts including the mouse, embryonated egg, and the rabbit.

Confirmation of the nature of the cytopathic effect may be obtained by the collodion technic of CHEATHAM et al. (1954). Typically, the infected cells are large and contain, one, occasionally two, inclusion bodies within the nucleus (Fig. 5). Some cells contain hyaline, eosinophilic or amphophilic cytoplasmic inclusions when stained with hematoxylin and eosin. These inclusions generally are less prominent, more variable, and less specific than the typical large nuclear inclusion.

4. Metabolism of Infected Cells

Because of the relatively slow progression of virus multiplication in tissue culture, pH differences between infected and control culture are not readily discernable and thus metabolic-inhibition is not a useful indicator of infection. A reduction in the pH may result in a deterioration of infected and uninfected cells and diminish the progression of the cytopathic effect. VONKA and BENYESH-MELNICK (1966) have noted that virus titers in 3 day old cultures are higher than in 5 day old cultures. Ten day old cultures, however, were not less sensitive to virus infection than 15 day old cultures. The metabolic explanation for these differences in sensitivity is not known.

F. Pathogenesis

1. Experimental Infection

Murine cytomegalovirus in sublethal doses can produce a focal acute hepatitis without residual cirrhosis or permanent liver damage (HENSON et al., 1966). In contrast, lethal doses result in cell necrosis and spreading foci of infection (McCORDOCK and SMITH, 1936). When virus is injected intraperitoneally it enters the liver via the portal veins. HENSON and associates (1966) have demonstrated an increase in Kupffer cells surrounding infected hepatocytes. The latter subsequently degenerate into acidophilic bodies. At that time both virus and interferon production diminishes in the liver. It is several days thereafter before the first neutralizing antibody is detectable. Irradiation does not interfere with virus interferon production but does abolish the Kupffer and inflammatory cell response. This results in expanding foci of necrosis, florid inclusions and death. These authors suggest that Kupffer cells, neutrophilic and lymphocytic inflammatory cells are important in the localization of infection in the liver and may be responsible for the transformation of infected hepatic parenchymal cells into acidophilic bodies. It is of interest that the X-irradiation sufficient to diminish the inflammatory response does not inhibit interferon production.

OSBORNE and MEDEARIS (1966) and GLASGOW et al. (1967) have noted diminished sensitivity of murine and human strains of CMV to exogenous interferon. These investigators have also noted that interferon produced by CMV-infected cells is less than that of several other viruses tested in replicate cultures. These data may have some relevance to the ability of CMV to persist in the host for unusually long periods of time (WELLER and HANSHAW, 1962). OSBORNE and

MEDEARIS have found that mice infected with CMV are more susceptible to NDV. The authors attribute this unusual susceptibility to the failure of CMV-infected mice to synthesize interferon in response to NDV infection.

Congenital defects or disease similar to that observed in the infant have not been demonstrated in experimental animals.

2. Natural Infection

The pathogenesis of human disease caused by CMV is obscure. It is apparent that infection must be passed from the mother to the fetus because the infant is born with signs of generalized disease. The mother usually tolerates the infection well. Most mothers giving birth to an infected infant do not have a history of a disease process that can be clearly attributed to CMV infection. MEDEARIS (1964) has demonstrated maternal viruria in 5 of 7 mothers of infants with CID. LeLong et al. (1960) have demonstrated CMV in placental tissue.

The marked difference in the response of the mother and the infant is not fully explained. The infant tends to remain infected longer, to excrete higher titers of virus (even in the presence of significant neutralizing antibody), and is more likely than the adult to have clinical symptoms related to his infection. Factors cited by MEDEARIS (1964) that may possibly account for this differing susceptibility include: the lessened integrity of the fetal blood brain barrier; the decreased phagocytic ability of fetal macrophages; the delayed immunoglobulin response; the greater number of viral receptor sites found on immature cells, and the deficient interferon capacity of the immature animal.

When a developing fetus is infected with CMV the effect is variable. No disease may result or there may be generalized, pantropic infection with inflammatory changes in many organs and eventual death. The factors determining variability in virulence are also poorly understood. Of 42 infants studied virologically by WELLER and HANSHAW (1962), MEDEARIS (1964), and HANSHAW (1966), 29 (69%) presented evidence of a fetal meningo-encephalitis as manifested by the subsequent failure of brain growth (microcephaly). There is a tendency for inflammation to occur in the lining ependymal cells of the ventricles, the posterior uvea, and the optic nerve. It is possible that a variety of developmental abnormalities may result if fetal infection occurs during the first trimester. Of the 42 infants cited above, 38 (90%) survived the fetal infection.

As in the experimental murine host, there is a tendency for virus to involve the liver in congenital and acquired human infections. Although the virus has been implicated as one cause of neonatal hepatitis in man (WELLER and HANSHAW, 1962; DANKS and associates, 1965; STERN and TUCKER, 1965) there is less certainty but growing suspicion in regard to the role of CMV in subacute and chronic hepatitis in older children and adults (HANSHAW et al., 1965; KLEMOLA and KAARIAINEN, 1965; LAMB and STERN, 1966).

In man there have been relatively few reports of CID in older individuals. However, CMV infection and disease have been associated with renal homotransplantation (KANICH and CRAIGHEAD, 1966; CRAIGHEAD et al., 1967; RIFKIND, 1965; HILL et al., 1964), leukemia (DUVALL et al., 1966; BODEY et al., 1965; HANSHAW and WELLER, 1961) and hypogammaglobulinemia (JACOX et al., 1965).

Thus, it appears that susceptibility to CMV is increased during fetal life and in a variety of conditions characterized by immunologic inadequacy. In the otherwise normal child or adult, however, the infection occurs with a variable, but usually mild host response.

Little is known about the communicability of CMV in experimental or natural hosts. Virus is present in the upper respiratory tract, saliva, and urine of man. HANSHAW et al. (1965) found that 28% of siblings and playmates of infected children also had demonstrable viruria.

G. Variation

There is no evidence to suggest that strains isolated from various sources or of different antigenic types exhibit variability in virulence or tissue tropism. Strain variability in terms of adaptability to tissue culture and in the progression of the CPE, has been observed. These characteristics may effect their suitability as complement-fixing antigens.

Table 1. *Prevalence of Cytomegalovirus Complement-fixing Antibody in Rochester, New York*[1]

Age	No. tested	No. positive
0— 5 months	48	17 (35.4%)
5—24 months	77	2 (2.6%)
2— 6 years	79	5 (6.3%)
6—10 years	201	18 (9.0%)
10—17 years	23	5 (21.7%)
17—40 years	159	60 (37.7%)
	587	107 (18.3%)

[1] Reproduced from HANSHAW (1966) with permission of the publisher.

H. Immunity
1. Active Immunity

Infection in utero results in the production of γM and frequently γA antibody (McCRACKEN and SHINEFIELD, 1965). These elevations may be useful as screening tests in the detection of intrauterine infections due to CMV or other agents including rubella, toxoplasmosis, or syphilis (STIEHM et al., 1966). HANSHAW et al. (1968) have recently demonstrated specific cytomegalovirus macroglobulin in congenitally infected infants using an indirect fluorescent antibody method. This test is useful in distinguishing active antibody formation from passive immunity of maternal origin. Women giving birth to infected infants frequently have high CMV complement-fixing antibodies (HANSHAW, 1966a). An infant receiving maternal γG (7S) antibodies transplacentally will have low or undetectable levels by five to six months of age. HANSHAW (1966a) has found that a CF antibody titer greater than 1:8 in a six month old infant usually results from active infection rather than passive transfer of maternal antibody. Because of the relative infrequency of CF antibody from 5—12 months of age (Table 1) the CF test can be useful in the diagnosis of infantile and congenital infection.

In older individuals acquiring infection for the first time, antibody usually rises over a three week period as in other viral infections (KLEMOLA et al., 1966). However, once acquired, the antibody is especially durable; an observation that may be due to the persistence of the CMV antigen in the host.

2. Passive Immunity

The infant of an antibody-positive woman is given protective passive immunity for approximately 4—6 months after birth. If contact with a CMV infected person occurs during these months when protective antibody is present, the infant may become naturally immunized and remain asymptomatic.

3. CMV Vaccine

There have been no experimental trials of a human CMV vaccine. If and when the need for an active immunizing agent is clearly established, the usually mild or asymptomatic nature of the acquired disease will be an advantageous feature in the preparation and possibly the acceptance of a vaccine. As in rubella, the possibility of transfer of infection from a vaccinated subject to a pregnant or particularly susceptible individual could be a significant problem in the development of a live virus vaccine.

J. Essential Clinical Features, Pathology and Diagnosis

1. Congenital and Infantile Infection

It may be difficult to determine if infection detected in the first year of life is congenital or acquired. One can assume that typical symptoms beginning in the first week of life of an infected infant are congenital in origin. There are infants who are asymptomatic during the neonatal period, who are

Table 2. *Manifestations of Cytomegalic Inclusion Disease in Infants Studied Virologically*

Clinical manifestations	Number with manifestation	
	WELLER and HANSHAW (1962)	MEDEARIS (1964)
	(17 cases)	(7 cases)
Hepatomegaly	17	7
Splenomegaly	17	6
Mental retardation	14	5
Microcephaly	14	7
Jaundice	11	7
Petechiae	9	7
Chorioretinitis	5	2
Cerebral calcifications	4	2

discovered to have extensive periventricular calcification (GUYTON et al., 1957) during the second month of life. This finding suggests an infection of some duration. Demonstration of viruria in a young infant may or may not be associated with some manifestation of CID of greater or lesser severity. Infants who are neurologically abnormal later may also be symptom-free in the early weeks of life (HANSHAW, 1966 b). More often, however, a significantly affected infant will have symptoms during the first days of life. It is not known if infants acquiring the infection after birth are at risk in terms of neurologic sequelae. Table 2 lists some of the clinical manifestations observed in 24 infants studied by WELLER and HANSHAW (1962) and MEDEARIS (1964). Abnormalities associated with infection may not necessarily be due to the direct effect of the virus. Thus, a 10 month old infant with a small head, cerebral palsy and serologic evidence of the infection may simply have two conditions coexisting but unrelated etiologically. These data are presented, however, to illustrate the frequency of associated oculo-cerebral abnormalities, particularly microcephaly, among CMV-infected infants. The extent and type of these abnormalities depend on the stage of fetal development when infection occurs as well as the severity of the infection.

There are preliminary data to suggest that the clinical manifestations of congenital CMV infection may be more varied than is presently appreciated and that any child with a reduced head circumference should be tested for CMV infection (HANSHAW, 1966a).

2. Infection Acquired after Infancy

CMV infection acquired after infancy may be symptomatic or asymptomatic. At this time CID may be associated with an underlying condition or therapy which interferes with the ability of the host to respond with specific antibody and perhaps nonspecifically with such host defenses as the elaboration of interferon.

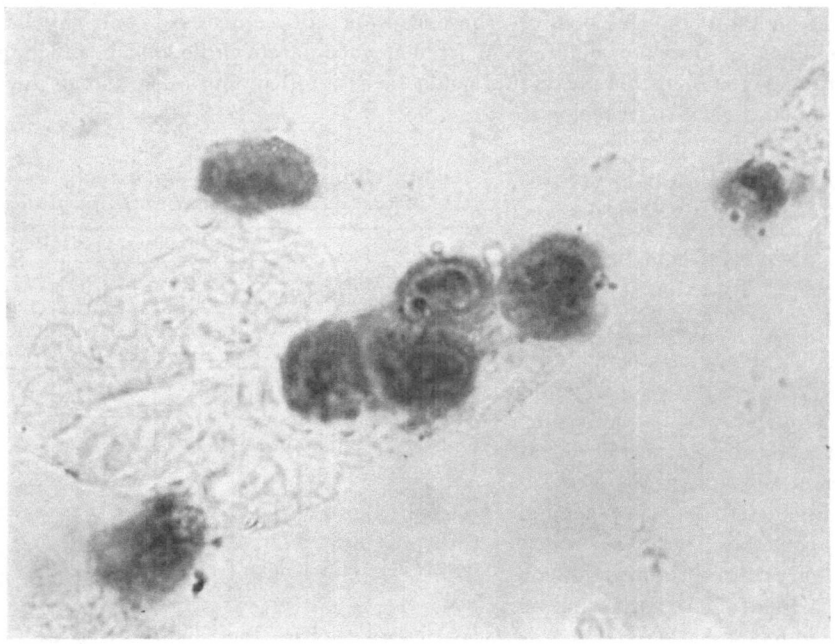

Fig. 6. Several inclusion-bearing cells in urinary sediment. Reproduced from HANSHAW (1964) with permission of the publisher.

Infection in a previously well child may result in enlargement of the liver and abnormal liver function tests (HANSHAW et al., 1965). In addition, an infectious mononucleosis-like blood picture (atypical lymphocytosis) with pyrexia, and malaise may be associated with liver dysfunction. Patients undergoing blood transfusions, especially multiple transfusions, have had a similar "mononucleosis-like disease" associated with a four-fold rise in CMV antibody titer (KAARIAINEN et al., 1966). ZUELZER et al. (1966) have described CMV infection in patients with acquired hemolytic anemia and postulate an etiologic relationship.

3. Pathology

FETTERMAN (1952) noted the pathognomonic cell of CMV infection in urine (Fig. 6). Typical cells have been described in the placenta, in gastric washings, and in most of the organs of the body. The most frequently involved tissues in congenital infection include brain, liver, kidney, adrenal gland, lung, and pancreas. In the acquired infection the lung is most often involved (NELSON and WYATT, 1959; SEIFERT and OEHME, 1957). Inclusions are also found occasionally in the liver, salivary glands, and gastrointestinal tract (WONG and WARNER, 1962). In man (WELLER and HANSHAW, 1962) and in mice (BRODSKY and ROWE, 1958) virus may be isolated from tissues devoid of inclusion-bearing cells.

In cytomegalovirus pneumonitis the inclusions are usually found in the epithelium of the alveoli (Fig. 1). Focal or diffuse interstitial pneumonitis occurs in the presence of a mononuclear infiltration in the alveolar walls. Pneumocystis carinii has been associated with CMV pneumonitis in debilitated infants (SEIFERT and OEHME, 1957) and in adults (SYMMERS, 1960), having disease known to increase susceptibility to infection. In the liver, inclusion-bearing cells are found most often in the small bile ducts (SMITH, 1959). WELLER and associates (1957) found giant cell formation in infants with CMV hepatitis. HANSHAW et al. (1965) have isolated virus from the liver of a 19 year old girl dying of "post-necrotic cirrhosis". Inclusion bearing cells were demonstrated in sections of the post-mortem liver tissue. LAMB and STERN (1966) have recently reported jaundice in an adult with "cytomegalovirus mononucleosis". More data are needed before the role of CMV in liver disease is fully understood.

When kidney involvement occurs, the proximal convoluted tubules are the most frequent site of inclusion formation. Infected cells exfoliate into the urine (Fig. 6) providing the basis for a useful diagnostic test. Less than half of the CMV infected infants studied by HANSHAW (1966b) have had such cells in the urine sediment. It is not known if renal CMV infection produces chronic kidney disease.

The pathology of CMV infection of the central nervous system has been limited to congenital disease and is described by DIEZEL (1954), COURVILLE (1961) and HAYMAKER et al. (1954). Infection produces an encephalitis with focal destruction of tissue, calcification, and severe inflammatory changes. Aplasia of the cerebellum or cerebral cortex may result from early gestational infection. The subependymal tissues are particularly prone to tissue destruction, giving the appearance of a periventricular distribution to radiologically demonstrable, calcified lesions. Although calcifications have been demonstrated in more peripheral areas of the brain, they usually occur in association with subependymal involvement.

The ability of the pathologist to demonstrate inclusions is dependent upon the stage and severity of the infection at the time the tissues are examined. Thus, an infant at 18 months of age dying from the sequellae of congenital CID may have no characteristic pathologic changes to suggest this diagnosis. WELLER and HANSHAW (1962) demonstrated inclusions in the liver of only two of four young infants with generalized CID. CMV, however, was isolated readily from liver biopsy material on all four patients.

4. Diagnosis

The diagnosis of CID requires virologic, serologic and clinical data. The presence of *infection* (not necessarily disease) can be established by: a) CMV isolation; b) demonstration of pathognomonic inclusions; c) a rising CMV antibody titer; d) detection of cytomegalovirus macroglobulin. An antibody titer of 1:8 in an infant 6—12 months of age is indicative of congenital or early infantile infection and is usually but not always associated with the symptoms of disease (HANSHAW, 1966).

Table 3. *Prevalence of Cytomegalovirus Complement-fixing Antibody in Different Childhood Populations*

Authors	Location	Age range tested	No. pos./ No. tested	Positive %
ROWE et al. (1958)	Washington, D.C.	6 months—15 yrs.	47/139	33
MENDEZ-CASHION et al. (1963)	San Juan, Puerto Rico	1—11 years	20/99	20
CARLSTROM (1965)	Stockholm, Sweden	7 months—15 yrs.	27/108	25
STERN and ELEK (1965)	London, England	6 months—15 yrs.	73/447	16
VACZI et al. (1965)	Debrecen, Hungary	4—14 years	33/120	28
DEIBEL et al. (1966)	Albany, New York	10 years	4/59	7
HANSHAW (1966)	Rochester, New York	5 months—17 yrs.	30/380	8
LI and HANSHAW (1967)	Migrants from Florida	1—13 years	22/42	52

Table 4. *Incidence of Cytomegalovirus Infection in Three Groups of Rochester Children*[1]

Group	Age range Year	No. of subjects tested	No. of positive tests
Well	0— 2	23	
	2— 6	77	1
	6—14	100	1
Totals		200	2 (1%)
Hospitalized	0— 2	28	
	2— 6	35	1
	6—14	37	
Totals		100	1 (1%)
Institutionalized	0— 2	1	1
	2— 6	19	3
	6—14	2	1
Totals		22	5 (23%)

[1] Reproduced from HANSHAW and associates (1965) with permission of the publisher.

Diagnosis based on clinical manifestations alone is subject to error because of the close resemblance of many of the clinical symptoms to congenital toxoplasmosis, congenital rubella, sepsis, etc. Periventricular calcification, while not absolutely pathognomonic of CMV infection *in utero*, is strongly suggestive of this disease.

In older infants and children, the distinction between infection and disease must be made with great care. Demonstration of infection becomes less significant with increasing age because of the greater likelihood of persistent asymptomatic infection and the increased probability of having acquired the infection from environmental contacts rather than from the mother *in utero*. Data as to the effect of primary infection after infancy are just becoming available. Future investigations may determine that persistent CMV infection is associated with chronic respiratory, hepatic or renal disease.

K. Epidemiology

There are several factors which determine the frequency and distribution of CMV infection.

Childhood populations vary in the prevalence of CMV complement-fixing antibody as determined by different investigators in different parts of the world (Table 3). LI and HANSHAW (1967) found CMV antibody among 22/42 children living in two New York State migrant labor camps. The crowding and poor sanitary conditions in these camps probably contribute to the rapid dissemination of virus in this population.

There is no evidence that climate is an important factor in the incidence of CMV infection. SEVER et al. (1963) found four-fold CMV antibody responses during the pregnancy of 5—6% of 198 pregnant women studied in Boston and Philadelphia. HANSHAW et al. (1968) have isolated cytomegalovirus from the urine of 6 of 280 infants cultured within four days of birth.

HANSHAW et al. (1965) studied the incidence of CMV infection among pediatric inpatients, children institutionalized for tuberculosis, and well children living at home. These data are summarized in Table 4. Data on antibody prevalence in the Rochester, New York area are given in Table 1 (HANSHAW, 1966).

Relatively little is known about the transmission of cytomegalovirus infection in man or animals. Since virus is present in the urine and saliva, it is likely that transmission is primarily through contact with these excretions. Theoretically, chronically infected infants could be important sources of spread of the virus in families and institutions. HANSHAW et al. (1965) observed a family of five children with active viruria, hepatomegaly and hepatic dysfunction. The oldest child was retarded and microcephalic, a circumstance suggesting congenital infection. It is postulated that this child continued to excrete virus after birth and infected his younger siblings. Mothers giving birth to infected infants may give a history of prolonged influenza-like illness or persistent cough during pregnancy. It is not established if these respiratory infections represent primary CMV disease in the mother. Once primary infection, viremia, and an antibody response occurs, it is considered unlikely that a mother would be able to transmit CMV infection to the fetus in subsequent pregnancies. This important point requires further study.

Although there are inadequate data available at this time on the actual risk involved, it would seem prudent to consider CMV as well as rubella-infected individuals potentially hazardous to pregnant women.

References

ANDREWS, C. H.: Classification of viruses of vertebrates. Advanc. Virus Res. 9, 271—296 (1962).

BECKER, P., J. L. MELNICK, and H. D. MAYOR: A morphological comparison between the developmental stages of herpes zoster and human cytomegalovirus. Exp. molec. Path. 4, 11—23 (1965).

BENYESH-MELNICK, M., V. VONKA, F. PROBSTMEYER, and I. WIMBERLY: Human cytomegalovirus: Properties of the complement-fixing antigen. J. Immunol. 99, 261—267 (1966).

BLACK, P. H., J. W. HARTLEY, and W. P. ROWE: Isolation of a cytomegalovirus from African green monkey. Proc. Soc. exp. Biol. (N.Y.) 112, 601—605 (1963).

BRODSKY, I., and W. P. ROWE: Chronic subclinical infection with mouse salivary gland virus. Proc. Soc. exp. Biol. (N.Y.) 99, 654—655 (1958).

CARLSTROM, G.: Virologic studies on cytomegalic inclusion disease. Acta paediat. (Uppsala) 54, 17—23 (1965).

CHEATHAM, W. J., cited in ENDERS, J. F., and T. C. PEEBLES: Propagation in tissue cultures of cytopathogenic agents from patients with measles. Proc. Soc. exp. Biol. (N.Y.) 86, 277—286 (1954).

COLE, R., and A. G. KUTTNER: A filtrable virus present in the submaxillary glands of guinea pigs. J. exp. Med. 44, 855—873 (1926).

COURVILLE, C.: Cerebral lesions in cytomegalic inclusion disease. Bull. Los Angeles neurol. Soc. 26, 9—21 (1961).

CRAIGHEAD, J. E., J. B. HANSHAW, and C. B. CARPENTER: Cytomegalovirus infection after renal allotransplantation. J. Amer. med. Ass. 201, 725—728 (1967).

CRAWFORD, L. V., and A. J. LEE: Discussion and preliminary reports. The nucleic acid of human CMV. Virology 23, 105—107 (1964).

DANKS, D. M., P. E. CAMPBELL, and J. F. CONNELLY: An aetiological study of neonatal jaundice in a children's hospital. Aust. paediat. J. 1, 193—201 (1965).

DEIBEL, R., R. J. FAIRLEY, and C. DUCHARME: Serological studies with the cytomegalovirus. Annual Report: New York State Department of Health (1965).

DIEZEL, P. B.: Mikrogyri infolge cerebraler Speicheldrüsen-Virusinfektion im Rahmen einer generalisierten Cytomegalie bei einem Säugling, zugleich ein Beitrag zur Theorie der Windungsbildung. Virchows Arch. path. Anat. 30, 109—130 (1954).

FARBER, S., and S. B. WOLBACH: Intranuclear and cytoplasmic inclusions ("protozoan-like bodies") in the salivary glands and other organs of infants. Amer. J. Path. 8, 123—126 (1932).

FETTERMAN, G. H.: New laboratory aid in clinical diagnosis of inclusion disease of infancy. Amer. J. clin. Path. 22, 424—425 (1952).

GLASGOW, L. A., J. B. HANSHAW, T. C. MERIGAN, and J. PETRALLI: Interferon and human cytomegaloviruses in vivo and in vitro. Proc. Soc. exp. Biol. (N.Y.) 125, 843—849 (1967).

GOODHEART, C. R., J. R. FILBERT, and R. M. McALLISTER: Human cytomegalovirus: Effect of 5-fluoro-2-deoxyuridine (FUDR) on viral synthesis and cytopathology. Virology 21, 530—532 (1963).

GOODPASTURE, E. W., and F. B. TALBOT: Concerning the nature of "protozoan-like" cells in certain lesions of infancy. Amer. J. Dis. Child. 21, 415—421 (1921).

GREEN, M.: Chemistry of the DNA viruses in "Viral and Rickettsial Infections of Man" (HORSFALL and TAMM eds.), Fourth Edition, p. 162, J. B. LIPPINCOTT Company, Philadelphia-Montreal (1965).

GUYTON, T. B., R. EHRLICH, W. A. BLANC, and M. H. BECKER: New observations in generalized cytomegalic inclusion disease of the newborn. Report of a case with chorioretinitis. New Engl. J. Med. 257, 803—807 (1957).

HAMPARIAN, V. V., M. R. HILLEMAN, and A. KETLER: Contributions to characterization and classification of animal viruses. Proc. Soc. exp. Biol. (N.Y.) 112, 1040—1050 (1963).

HANSHAW, J. B.: The clinical significance of cytomegalovirus infection. Postgrad. Med.
    35, 472—480 (1964).
HANSHAW, J. B.: Cytomegalovirus complement-fixing antibody in microcephaly.
    New Engl. J. Med. 275, 476—479 (1966 a).
HANSHAW, J. B.: Congenital and acquired cytomegalovirus infection. Pediat. Clin.
    N. Amer. 13, 279—293 (1966 b).
HANSHAW, J. B., H. J. STEINFELD, and C. J. WHITE: Fluorescent antibody test for
    cytomegalovirus macroglobulin. To be published (1968).
HANSHAW, J. B., R. F. BETTS, G. SIMON, and R. C. BOYNTON: Acquired cytomegalo-
    virus infection: Association with hepatomegaly and abnormal liver function tests.
    New Engl. J. Med. 272, 602—609 (1965).
HANSHAW, J. B., and T. H. WELLER: Urinary excretion of cytomegaloviruses by
    children with generalized neoplastic disease. Correlation with clinical and histo-
    pathologic observations. J. Pediat. 58, 305—311 (1961).
HARTLEY, J. W., W. P. ROWE, and R. J. HUEBNER: Serial propagation of the guinea
    pig salivary gland virus in tissue culture. Proc. Soc. exp. Biol. (N.Y.) 96, 281—
    285 (1957).
HAYMAKER, W., B. R. GIRDANY, J. STEPHENS, R. D. LILLIE, and G. H. FETTERMAN:
    Cerebral involvement with advanced periventricular calcification in generalized
    cytomegalic inclusion disease in the newborn: A clinicopathological report of a
    case diagnosed during life. J. Neuropath. exp. Neurol. 13, 562—586 (1954).
HENSON, D., R. D. SMITH, J. GEHRKE: Non-fatal mouse cytomegalovirus hepatitis.
    Amer. J. Path. 49, 871—888 (1966).
HILL, R. B., Jr., D. T. ROWLANDS, Jr., and D. RIFKINDS: Infectious pulmonary
    disease in patients receiving immunosuppressive therapy for organ transplantation.
    New Engl. J. Med. 271, 1021—1027 (1964).
HORN, R. G., S. S. SPICER, and B. K. WETZEL: Phagocytosis of bacteria by heterophil
    leukocytes. Acid and alkaline phosphatase cytochemistry. Amer. J. Path. 45,
    327—335 (1964).
JACOX, R. F., E. S. MONGAN, J. B. HANSHAW, and J. P. LEDDY: Hypogamma-
    globulinemia with thymoma and probably pulmonary infection with cytomegalo-
    virus. New Engl. J. Med. 271, 1091—1096 (1965).
JESIONEK, A., und B. KIOLEMENOGLOU: Über einen Befund von protozoenartigen
    Gebilden in den Organen eines heriditärluetischen Fötus. Münch. med. Wschr. 51,
    1905—1907 (1904).
KAARIAINEN, L., E. KLEMOLA, and J. PALDHEIMO: Rise of cytomegalovirus anti-
    bodies in an infectious, mononucleosis-like syndrome after transfusion. Brit. med.
    J. 1, 1270—1272 (1966).
KANICH, R. E., and J. E. CRAIGHEAD: Cytomegalovirus infection and cytomegalic
    inclusion disease in renal homotransplant recipients. Amer. J. Med. 40, 874—882
    (1966).
KLEMOLA, E., and L. KAARIAINEN: Cytomegalovirus as a possible cause of a disease
    resembling infectious mononucleosis. Brit. med. J. 2, 1099—1102 (1965).
KLEMOLA, E., I. SALMI, L. KAARIAINEN, and A. KOIVUNIEMI: Hepatosplenomegaly
    after "cytomegalovirus mononucleosis" in a child. Ann. Paediat. Fenn. 12, 39—
    42 (1966).
LAMB, S. G., and H. STERN: Cytomegalovirus mononucleosis with jaundice as presen-
    ting sign. Lancet 2, 1003—1006 (1966).
LE LONG, M., F. LEPAGE, LE-TAN-VINK, P. TOURNIER, and C. CHANY: The virus of
    cytomegalic inclusion disease: its isolation in 2 cases. Survival of the patient
    without sequelae in one of these cases. Presence of inclusions in the placenta of a
    third case. Arch. franç. Pédiat. 17, 437—450 (1960).
LI, F. P., and J. B. HANSHAW: Cytomegalovirus infection among migrant children.
    Amer. J. Epidem. 86, 137—141 (1967).
LIPSCHÜTZ, B.: Untersuchungen über die Ätiologie der Krankheiten der Herpes-
    gruppe (Herpes zoster, Herpes genitalis, Herpes febrilis). Arch. Derm. Syph.
    (Berl.) 136, 428—482 (1921).

LÖWENSTEIN, C.: Über protozoenartige Gebilde in den Organen von Kindern. Zbl. allg. Path. path. Anat. **18**, 513—518 (1907).

LUSE, S. A., and M. G. SMITH: Electron microscopy of salivary gland viruses. J. exp. Med. **107**, 623—632 (1958).

LWOFF, A., R. HORNE, and P. TOURNIER: A system of viruses. Cold Spr. Harb. Symp. quant. Biol. **27**, 51—55 (1962).

MCALLISTER, R. M., R. M. STRAW, J. E. FILBERT, and C. R. GOODHEART: Human cytomegalovirus. Cytochemical observations of intracellular lesion development correlated with viral synthesis and release. Virology **19**, 521—531 (1963).

MCCORDOCK, H. A., and M. G. SMITH: The visceral lesions produced in mice by the salivary gland virus of mice. J. exp. Med. **63**, 303—310 (1936).

MCCRACKEN, G. H., Jr., and H. R. SHINEFIELD: Immunoglobulin concentrations in newborn infants with congenital cytomegalic inclusion disease. Pediatrics **36**, 933—937 (1965).

MCGAVRAN, M. H., and M. G. SMITH: Ultrastructural, cytochemical and microchemical observations on cytomegalovirus infection of human cells in tissue culture. Exp. molec. Path. **4**, 1—10 (1965).

MEDEARIS, D. N., Jr.: Observations concerning human cytomegalovirus infection and disease. Bull. Johns Hopk. Hosp. **114**, 181—211 (1964a).

MEDEARIS, D. N., Jr.: Viral infections during pregnancy and abnormal human development. Amer. J. Obstet. Gynec. **90**, 1140—1148 (1964b).

MENDEZ-CASHION, D., M. VALCARCEL, R. R. DE ARELLANO, and W. P. ROWE: Salivary gland virus antibodies in Puerto Rico. Bol. Asoc. med. P. Rico **55**, 447—455 (1963).

NELSON, J. S., and J. P. WYATT: Salivary gland virus disease. Medicine **38**, 223—241 (1959).

NIVEN, J. S. F.: Fluorescence microscopy of nucleic acid changes in virus-infected cells. Ann. N.Y. Acad. Sci. **81**, 84—88 (1959).

OSBORN, J. E., and D. N. MEDEARIS, Jr.: Suppression of interferon and antibody and multiplication of Newcastle disease virus in cytomegalovirus infected mice. Proc. Soc. exp. Biol. (N.Y.) **124**, 347—353 (1967).

PATRIZI, G., J. N. MIDDLEKAMP, J. C. HERWEG, and H. K. THORNTON: Human cytomegalovirus, electron microscopy of a primary viral isolate. J. Lab. clin. Med. **65**, 825—838 (1965).

PLUMMER, G., and M. BENYESH-MELNICK: A plaque reduction neutralization test for human cytomegalovirus. Proc. Soc. exp. Biol. (N.Y.) **117**, 145—150 (1964).

PLUMMER, G., and B. LEWIS: Thermoinactivation of herpes simplex virus and cytomegalovirus. J. Bact. **89**, 671—674 (1965).

RIBBERT, H.: Über protozoenartige Zellen in der Niere eines syphilitischen Neugeborenen und in der Parotis von Kindern. Zbl. allg. Path. **15**, 945—948 (1904).

RIFKIND, D.: Cytomegalovirus infection after renal transplantation. Arch. intern. Med. **116**, 554—558 (1965).

ROWE, W. P., J. W. HARTLEY, S. WATERMAN, H. C. TURNER, and R. J. HUEBNER: Cytopathogenic agent resembling human salivary gland virus recovered from tissue cultures of human adenoids. Proc. Soc. exp. Biol. (N.Y.) **92**, 418—424 (1956).

RUEBNER, B. H., T. HIRAND, R. SLUSSER, J. OSBORN, and D. N. MEDEARIS, Jr.: Cytomegalovirus infection: Viral ultrastructure with particular reference to the relationship of lysosomes to cytoplasmic inclusions. Amer. J. Path. **48**, 971—989 (1966).

RUEBNER, B. H., K. MIYAI, R. J. SHISSER, P. WEDEMEYER, and D. N. MEDEARIS, Jr.: Mouse cytomegalovirus infection; an electron microscope study of hepatic parenchymal cells. Amer. J. Path. **44**, 799—821 (1964).

SEIFERT, G., and J. OEHME: Pathologie und Klinik der Cytomegalie. Verl. Georg Thieme, Leipzig, 1957.

SEVER, J. L., R. J. HUEBNER, G. A. CASTELLANO, and J. A. BELL: Serologic diagnosis "en masse" with multiple antigens. Amer. Rev. resp. Dis. **88** (Supp.) 342—359 (1962).

Smith, K. O.: Physical and biological observations of herpesvirus. J. Bact. 86, 999—1009 (1963).

Smith, K. O., and L. E. Rasmussen: Morphology of cytomegalovirus (Salivary gland virus). J. Bact. 85, 1319—1325 (1963).

Smith, M. G.: Propagation of the salivary gland virus of the mouse in tissue cultures. Proc. Soc. exp. Biol. (N.Y.) 86, 435—440 (1954).

Smith, M. G.: Propagation in tissue cultures of a cytopathogenic virus from human salivary gland virus disease. Proc. Soc. exp. Biol. (N.Y.) 92, 424—430 (1956).

Smith, M. G.: The salivary gland viruses of man and animals. Progr. med. Virol. 2, 171—202 (1959).

Steihm, E. R., A. J. Ammann, and J. D. Cherry: Elevated cord macroglobulins in the diagnosis of intrauterine infections. New Engl. J. Med. 275, 971—977 (1966).

Stern, H., and S. D. Elek: The incidence of infection with cytomegalovirus in a normal population: A serological study in greater London. J. Hyg. (Lond.) 63, 79—87 (1965).

Stern, H., and I. Friedmann: Intranuclear formation of cytomegalic inclusion disease virus. Nature (Lond.) 188, 768—770 (1960).

Symmers, W. St. C.: Generalized cytomegalic inclusion-body disease associated with pneumonitis pneumonia in adults. J. clin. Path. 13, 1—21 (1960).

Tyzzer, E. E.: The histology of the skin lesions in varicella. J. med. Res. 14, 361—392 (1906).

Vaczi, L. et al.: Isolation of cytomegalovirus and incidence of complement-fixing antibodies against cytomegalovirus in different age groups. Acta microbiol. Acad. Sci. hung. 12, 115—129 (1965).

Vonka, V., and M. Benyesh-Melnick: Thermoinactivation of human cytomegalovirus. J. Bact. 91, 221—226 (1966a).

Vonka, V., and M. Benyesh-Melnick: Interactions of human cytomegalovirus with human fibroblasts. J. Bact. 91, 213—220 (1966b).

Watson, D. H., and P. Wildy: Some serological properties of herpes virus particles studied with the electron microscope. Virology 21, 100—111 (1963).

Weller, T. H.: Cytomegalovirus, in "Viral and Rickettsial Infections of Man" (Horsfall and Tamm, eds.), Fourth ed., p. 926—931. J. B. Lippincott Company, Philadelphia-Montreal, 1965.

Weller, T. H., and J. B. Hanshaw: Virologic and clinical observations on cytomegalic inclusion disease. New Engl. J. Med. 266, 1233—1244 (1962).

Weller, T. H., J. B. Hanshaw, and D. E. Scott: Serologic differentiation of viruses responsible for cytomegalic inclusion disease. Virology 12, 130—132 (1960).

Weller, T. H., J. E. Macauley, J. M. Craig, and P. Wirth: Isolation of intranuclear inclusion producing agents from infants illnesses resembling cytomegalic inclusion disease. Proc. Soc. exp. Biol. (N.Y.) 94, 4—12 (1957).

Weller, T. H., and W. P. Rowe: The human cytomegaloviruses in "Diagnostic Procedures for Viral and Rickettsial Diseases" (Lennette and Schmidt, eds.), Third edition, p. 707, American Public Health Association, Inc., New York, 1964.

Wong, T., and N. E. Warner: Cytomegalic inclusion disease in adults. Report of 14 cases with review of literature. Arch. Path. 74, 403—422 (1962).

Zuelzer, W. W., C. S. Stulberg, R. H. Page, J. Tervya, and A. J. Brough: Etiology and pathogenesis of acquired hemolytic anemia. Anemia 6, 438—461 (1966).

Rinderpest Virus

By

Walter Plowright

East African Veterinary Research Organization, Muguga
P.O. Kabete, Kenya, E. Africa

With 10 Figures

Table of Contents

I. Introduction and History

Prior to and during the first few centuries of the Christian era pestilences were recorded which afflicted indiscriminately man, his domestic animals and the wild fauna. As LECLAINCHE (1936) has noted, however, a sequence of events was quite regularly described in which at first dogs, then birds, horses, cattle, or wild animals and finally man perished in hordes; it was not until the great European epizootic of A.D. 376—386 that rinderpest finally came to be recognised as a distinct clinical entity, primarily affecting cattle; *the* cattle plague had arrived.

Until modern times rinderpest was the most important and devastating disease of cattle, being an almost inevitable concomitant of the wars which ravaged Europe at intervals, sending armies on the march with their trek oxen (GAMGEE, 1866; DIECKERHOFF, 1890). The infection often came to Europe from the East, allegedly the Caspian basin or the Russian and Hungarian steppes and extending along the valley of the Danube. In the 18th century alone it is estimated that about 200 million cattle in Europe fell victim to the disease (CURASSON, 1932; 1942). In the 19th century the development of an international trade in live cattle provided further opportunities for its rapid, wide dissemination and led to such catastrophic outbreaks as the 1865 epizootic in Great Britain, introduced with cattle transported from Revel by sea.

The need to combat rinderpest was undoubtedly an important factor in the establishment of the first veterinary schools in Europe, beginning with Alfort in 1762, as well as in the decision to convene the first International Veterinary Congress at Hamburg in July, 1863. More recently it was responsible in large measure for the foundation of the *Office Internationale des Épizooties* in Paris (VITTOZ, 1963), following the introduction of the disease to Belgium in 1920, by cattle trans-shipped at Antwerp *en route* from India to Brazil (ANON, 1920).

By 1930, Europe, with the exception of parts of Turkey, was free of the disease, thanks largely to the rigorous application of control measures, including slaughter of affected animals and the use of antiserum. The only outbreaks since that time have been small and confined to the primary foci, as in Rome in 1951 (CILLI et al., 1951) and Trieste in 1954 (BOLDRINI, 1954). Rinderpest has, however, continued to persist in Asia, especially in the south and south-east, as well as in the Indian sub-continent; in 1959, for example, there were about 8,000 outbreaks in India alone (ANON, 1965). The only outbreak recorded in the New World had its origin in Indian cattle imported into Brazil in 1920 (ROBERTS, 1921) and Australasia has only been infected once, in 1923 (ROBERTSON, 1924; WESTON, 1924).

In Africa the infection gained repeated access to the Nile Valley from Europe or the Near East and, according to CURASSON (1932; 1942), it probably spread as far as the Sudan and West Africa in 1805, 1828 and 1865. The most spectacular epizootic of recent times was, however, that which is reputed to have been introduced with Indian cattle to Somaliland during the Italian invasion of Ethiopia in 1889 (LUGARD, 1893; HUTCHEON, 1902). It had spread westwards across the present territories of Kenya and Uganda by 1890 and as far south as Lake Nyasa by 1892; it was officially reported in Bulawayo (Rhodesia) in 1896 and by 1897 had spread throughout the Transvaal and Orange Free State to reach the Cape Colony, Angola and S.W. Africa.

Cattle and susceptible game species (SIMON, 1962) were decimated, the losses of cattle in Rhodesia alone being 1.5 million; in the Cape Colony 1.3 million; in Bechuanaland and the Transvaal 1.0 million each (CURASSON, 1932; 1942). The mortality was often in excess of 95%, sometimes "limited to perhaps 90%" (LUGARD, 1893; HENNING, 1956). The disease also spread westwards as far as the west African colonies of France and Great Britain, reaching the Cameroons by 1891 and causing a similar mortality as in the south of the continent (CURASSON, 1932; 1942). Rinderpest was eliminated from the Cape Colony by 1903 and soon afterwards from the rest of South Africa, it remained enzootic, however, and has continued to cause recurrent epizootics in East, North-East, Equatorial and West Africa.

Several review articles on rinderpest have appeared in recent years, including an excellent and comprehensive synopsis by SCOTT (1964) and a more selective exposition of recent work (ANON, 1966a). Both of these have served as frequent sources for the present chapter, as have also the invaluable monographs of CURASSON (1932; 1942). No attempt has been made to prepare a complete bibliography but sufficient references have been given to provide an entrée to the very copious literature, especially that of the last 25—30 years.

II. Classification and Nomenclature

The classification of rinderpest virus during the last 20 years has passed through the usual transition phases. Thus, in the schema proposed by HOLMES (1948) it was placed in the genus *Tortor* because of its wide range of tissue tropisms. With the demonstration of a distinctive cytopathogenicity for rinderpest virus *in vitro* it was suggested (PLOWRIGHT and FERRIS, 1957) that the agent should be classified in the group III b of ENDERS, *i.e.* viruses causing large syncytial aggregates, with common cytoplasmic matrix and many nuclei; this group included measles, monkey kidney agents and, presumably, mumps (ENDERS, 1954; HENLE and DEINHARDT, 1955).

At about the same time the first reports appeared of an immunological relationship between rinderpest and canine distemper viruses (POLDING and SIMPSON, 1957), the latter already being linked serologically with the virus of human measles (ADAMS and IMAGAWA, 1957; CARLSTROM, 1957). Resemblances in the natural pathology of measles (infant) pneumonitis and canine distemper had been observed some years earlier (PINKERTON et al., 1945; ADAMS, 1953) when THIERY (1956) demonstrated for the first time that syncytia and specific viral inclusions, comparable to those seen in measles and distemper, were found in the epithelial and lymphoid tissues of rinderpest-infected cattle.

Finally, it was reported that adult human sera contained neutralising antibody for rinderpest virus (PLOWRIGHT and FERRIS, 1959a) and the stage was set for postulating that the causal viruses of three major animal plagues — measles, canine distemper and rinderpest — constituted a closely-related group (KOPROWSKI, 1958; IMAGAWA et al., 1960; WARREN, 1960).

COOPER (1961) was influenced by the large size of measles virus and its production of intranuclear inclusions to suggest tentatively that the measles-rinderpest-distemper (MRD) group should be classified with herpes viruses, as lipid-containing (ether-sensitive) deoxyviruses. However, the morphology of the mature virus particles, as seen in electron-micrographs after negative-contrast staining, soon rendered this hypothesis untenable. Measles (WATERSON et al., 1961), rinderpest (PLOWRIGHT et al., 1962) and distemper virions (CRUICKSHANK et al., 1962) were shown to have a structure which was virtually identical with that of Newcastle disease and other larger myxoviruses (WATERSON, 1962).

This finding provided strong presumptive evidence that the MRD sub-group would prove, like the myxoviruses, to possess ribonucleic acid (RNA) and not deoxyribonucleic acid (DNA). Hence, when LWOFF et al. (1962) proposed their "System of Viruses", they provisionally included these agents among the larger myxoviruses, i.e. enveloped riboviruses, having a helical symmetry and a nucleo-capsid of about 1700Å diameter. The Provisional Committee for the Nomenclature of Viruses (ANDREWES, 1965) proposed that the Family name of this group should be *Paramyxoviridae* but omitted to suggest a generic name for the clearly defined MRD sub-group.

III. Properties of the Virus
A. Morphology

Rinderpest was first shown to be filterable by NICOLLE and ADIL BEY (1902); prior to that time there had been difficulty in demonstrating its filterability, largely because the majority of the infectivity in the blood of infected cattle was closely associated with the leukocytes and was not free in the plasma or serum (see CURASSON, 1932; 1942). NICOLLE and ADIL BEY (1902) used BERKEFELD filters of reduced thickness to pass brain emulsions and diarrheic fluids; they also produced peritoneal exudates which were filtered successfully through normal BERKEFELD candles.

The first estimates of the particle size of rinderpest virus were obtained by CARMICHAEL and HUGHES (SIMMONS, 1941) who filtered bovine tissue extracts through graded collodion membranes and found the limiting average pore diameter (APD) to be 1670Å. Applying the factor of 0.64, as recommended by BLACK (1958), this corresponds with a minimal particle diameter of 1067Å, whereas CARMICHAEL and HUGHES calculated the size of the virus to be 840—1260Å.

PLOWRIGHT et al. (1962) reported on the morphology of virus particles as seen in tissue culture fluids infected with the virulent Kabete '0' (RBOK) strain; additional observations (WATERSON — unpublished) were made on two further strains of virus, RBT/4 and RBK/2, recently isolated from field outbreaks of mild rinderpest (PLOWRIGHT, 1963a). Preparations were concentrated about 100-fold by centrifugation for one hour at approximately 35,000 g and examined electron-micrographically after formalin treatment and negative staining with phosphotungstate at pH 7.0 (BRENNER and HORNE, 1959). The majority of particles were spherical or ovoid with a maximum diameter of 1200—3000Å; some were as large as 7500Å with a more irregular outline.

In preparations of RBOK virus, of the 91st calf kidney passage, there were also filaments or elongated forms with a width of 300—500Å and length up to 10,000Å; similar forms were seen less frequently in preparations of the RBT/4 strain. The viral nature of these filaments was evidenced (1) by the presence of outer projections similar to those of ordinary particles; (2) by their occurrence in a form connected to normal particles and (3) by their similarity to the filamentous forms of influenza or the Blacksburg strain of Newcastle disease virus (NDV). A few ring forms similar to those produced by several strains of NDV were also seen. Most particles were bounded externally by a conspicuous membrane, provided on the outside with a layer of projections, up to 90Å long and best seen in profile; these structures were lacking over parts of the surface of some particles. According to PROVOST et al. (1965b) the outer membrane with its peripheral spikes was acquired as the nucleocapsid passed through the cell membrane by a process of microvillus formation and "budding", analogous to that observed with other large myxoviruses, such as parainfluenza 3 (RECZKO and BÖGEL, 1962).

It was found that numerous particles began to disrupt spontaneously, as shown by breaks in the outer membrane, penetration of the particles by the phosphotungstate and release of segments of the internal component (PLOWRIGHT et al., 1962). The latter closely resembled the inner component of NDV, parainfluenza viruses, mumps and other members of the MRD group. It consisted of a flexible filament, normally coiled in the centre of the particle like a ball of wool; the diameter was about 175Å and the length of the filamentous structure in a single particle attained 25—35,000 Å. The surface of the filament was composed of helically-arranged subunits, giving a serrated or herring-bone appearance with a periodicity of 50—60Å. These figures are generally consistent with similar data for the internal component of measles virus and NDV, as summarised by WATERSON (1965); he calculated the average length of the nucleo-capsid to be about 40,000Å with 10,000 nucleotides or 3.3×10^6 molecular weight units.

B. Physico-chemical Structure

There is no direct evidence on the physico-chemical constitution of rinderpest virus and the only sources of information are therefore indirect, primarily from the work on the structurally-similar virus of measles or, secondly, from the application of cytochemical techniques to infected cells and the use of metabolic inhibitors. The two latter approaches will be considered first.

1. Application of Cytochemical Techniques

LIESS (1964) reported that the cytoplasmic inclusions which developed in infected HeLa or calf kidney monolayers were basophilic or neutrophilic in preparations stained by the MAY-GRÜNWALD:GIEMSA method (JACOBSON and WEBB, 1952) and these inclusions corresponded with the sites of specific immunofluorescence; in the early stages of infection the basophilic material could be completely removed by ribonuclease (RNase) treatment. The cytoplasmic inclusions never contained Feulgen-positive material and the use of RNase in conjunction with acridine orange or methyl-green-pyronin staining showed that only RNA was incorporated into these structures (LIESS, 1964); these results were confirmed in bovine embryonic kidney cells by PROVOST et al. (1965b). The intranuclear inclusions, which were also prominent in cells infected with some isolates of rinderpest virus, similarly failed to show any evidence of deoxyribonucleic acid at any stage of their development (LIESS, 1964).

2. Use of Metabolic Inhibitors

PLOWRIGHT (1962a) failed to observe any inhibitory effect on viral multiplication when 5-fluoro-deoxyuridine was incorporated, at levels up to 10^{-5} M, in the medium of calf kidney monolayers infected with rinderpest virus; unfortunately he did not include known DNA and RNA viruses as controls. PROVOST et al. (1965) found that 5-iodo-deoxyuridine, at 10^{-5} M concentration, had no effect on the growth of rinderpest virus in bovine embryonic kidney cells but did completely suppress vaccinia virus multiplication; the latter inhibitory effect was reversed by 10^{-5} M thymidine.

3. Resemblance to Measles Virus and Other Large Myxoviruses

Because of the close morphological resemblances and immunological relationships between rinderpest, measles and distemper viruses it seems reasonable to suppose that all these agents are physico-chemically of a similar structure. Furthermore, it would be surprising if they differed importantly from the larger myxoviruses such as NDV or parainfluenza viruses.

Investigation of the structure of measles virus has been greatly facilitated by the demonstration of haemagglutinins (PERIÉS and CHANY, 1960) and the use of Tween-ether treatment to release these from the nucleocapsid material (NORRBY, 1962a; WATERSON et al., 1963). The general conclusions derived from these studies (WATERSON, 1965), are that the haemagglutinating and, probably, the immunising activity of measles virus preparations are attributable to envelope materials, possibly of a glycoprotein nature and forming, after Tween-ether treatment, rosette-like forms similar to those of NDV (ROTT and SCHÄFER, 1961). The nucleocapsid, which is almost certainly ribonucleoprotein (LAM and ATHERTON, 1963; LEVINE and OLSON, 1963; NORRBY et al., 1964), consists of filamentous structures comparable to those of the soluble NP antigen of NDV (ROTT et al., 1963) and has no haemagglutinating activity.

As will be shown later, all efforts to demonstrate haemagglutinins for rinderpest virus have so far failed and it is, therefore, much more difficult to perform experiments with this virus on lines similar to those mentioned above. Nevertheless, the existence of envelope materials comparable to those of measles virus can be confidently inferred from the observations that rinderpest antisera inhibit measles haemagglutinin to high titre (WATERSON et al., 1963; BÖGEL et al., 1964) and the receptor sites for measles haemagglutinin on monkey *(E. patas)* erythrocytes can be blocked by prior treatment with rinderpest antigens (ANON, 1966b).

C. Antigenic Structure

Like the other members of the MRD group, strains of rinderpest virus show an immunological homogeneity which probably reflects a basically similar structure of the surface antigens. It is not known, however, which, if any, of the antigens demonstrable by *in vitro* serological tests are associated with the production of immunity or, in fact, how any of them are related to structures in the virion or virus-specific products of infected cells. The tests which have been applied include the complement fixation (C/F) and agar gel diffusion precipitation (AGDP) reactions which are described below:

1. Complement-fixing Antigens

Since 1916 these have been investigated intensively by Japanese workers and especially NAKAMURA and his associates (NAKAMURA, 1958). Initially they employed a system of heated lymph node extracts and unheated hyperimmune ox serum but they later used dried, unheated, tissue antigens to demonstrate antibodies in the sera of convalescent cattle, sheep and rabbits (NAKAMURA, 1951). The heated antigens were prepared by boiling 10% saline suspensions of tissue, for 30—60 minutes, and then clarifying at 3—4,000 r.p.m. for 15—30 minutes; such extracts were apparently unaffected by 2 cycles of freezing (—20°C) and thawing but were rapidly destroyed in putrefying tissue. They did not react well with convalescent antibody and other antigens, stable at room temperature, were therefore prepared by prolonged drying of lymph node tissues under vacuum over $CaCl_2$ and extracting the dessicated powder immediately before use with $N/50$ NaOH.

A similar method was used by WHITE (1958b) and by SCOTT and BROWN (1961) for the production of a dried antigen giving a single line of precipitation in AGDP tests; the latter authors suggested that this precipitin was probably identical with "the complement-fixing antigen" and stated that it was heat labile and readily inactivated by ether and methyl alcohol. WALKER et al. (1946a) prepared complement-fixing "soluble" antigens, freed of infectivity by centrifuging cattle spleen extracts at 15,000 r.p.m. for 1 hour and BOULANGER (1957a, b) used 10,000 r.p.m. for one hour to clarify saline extracts of acetone-ether extracted spleen from rabbits and cattle; his technique followed essentially that of CASALS et al. (1951). BOULANGER (1957a) also attempted unsuccessfully to deposit the antigen from rabbit spleen extracts by spinning at 40,000 r.p.m. for one hour.

STONE and MOULTON (1961) used a simplified technique for the production of complement-fixing antigens from infected cattle tissues, the primary objective, as in all the previously-quoted work, being to devise rapid diagnostic procedures, rather than to characterise the antigen(s) involved. They clarified crude lymph node extracts at low-speed and then a second time at 9—10,000 g for 30 minutes; the clear supernatant was also used as an antigen in both C/F and AGDP tests employing hyperimmune rabbit serum (STONE, 1960). STONE and DELAY (1961) showed that the titre of complement-fixing antigen in their preparations was not affected by treatment with as much as 1.0% of β-propiolactone (BPL) for 30 minutes at pH 8.0, although infectivity was destroyed by 0.4% BPL in the same time. Heating at 56°C for 30 minutes reduced the C/F titre of one preparation from 1/560 to 1/300 (STONE, 1960). WHITE and COWAN (1962) separated the "soluble" complement-fixing antigen from the infectious particles in suspensions of infected rabbit lymph nodes by centrifuging at 103,000 g for 120 minutes. The antigen in the supernatant was inactivated by heating at 56°C for 30 minutes, precipitated by ammonium sulphate and eluted from diethylaminoethyl (DEAE) cellulose columns in a manner suggestive of a protein.

It will be seen that there are considerable differences in the reported properties of the complement-fixing antigens derived from rinderpest infected tissues. For example, those described by NAKAMURA (1958) resist boiling for 30—60 minutes, whereas heating at 56°C for 30 minutes reduced the C/F antigen titre for STONE (1960) and abolished AGDP activity for WHITE and COWAN (1962). The work of BOULANGER (1957a, b) demonstrated an antigen which was not affected by lipid solvents (acetone and ether) whereas SCOTT and BROWN (1961) referred to ready inactivation by ether and methyl alcohol. It has been suggested (ANON, 1966a) that the thermostable antigen is glycoprotein in nature, comparable to the haemagglutinin of measles virus, or a carbohydrate-lipid complex, but there is no evidence to support these contentions. Further work is evidently necessary, especially with a view to separating and characteristing the antigens involved and elucidating their relationship to virus synthesis and structure. Some preliminary steps have been taken by applying the AGDP technique.

2. Agar-gel Diffusion Precipitating Antigens

WHITE (1958a, b) was the first to demonstrate a single virus-specific precipitinogen in crude extracts of infected bovine lymph nodes; he used the Ouchterlony technique, and hyperimmune rabbit serum, remarking that he had failed to obtain similar results with convalescent cattle and rabbit sera. STONE (1960) found two precipitinogens in cattle lymph node extracts, of which the faster migrating one was completely inactivated by heating at 56°C for 30 minutes; starch-block electrophoresis also showed that particles of two different electrophoretic mobilities were concerned. WHITE and COWAN (1962) separated a single AGDP antigen from the infectivity by centrifuging at 103,000 g for 120 minutes and found that it was thermolabile.

More recently, ISHII et al. (1964) demonstrated three antigens in preparations extracted with veronal-buffered saline from dried lymph node material (NAKAMURA, 1951); two of these were resistant to heating at 100°C for 30 minutes while the third, a more slowly migrating one, was completely inactivated. The serum used by these workers was prepared by hyperimmunisation of cattle, not rabbits, which had been employing by previous investigators. ISHII et al. (1964) demonstrated that the antibodies could be absorbed by prior treatment with the appropriate antigens and that the lines were therefore specific; rabbit lymph node extracts contained only the fastest-moving, heat-stable antigen and not the other two. ISHII et al. (1964) further found that all the three antigens described remained in the supernatant after centrifuging at 40,000 r.p.m. for 1 hour and were not present in the deposit. SCOTT and BROWN (1961) mentioned that "on occasion" two bands of precipitate occurred with crude lymph node extracts diffused against hyperimmune rabbit serum but usually only one was present.

The explanation for the different number of antigens demonstrated by different workers has still to be elucidated; it could evidently depend on the source and mode of production of the antiserum, the virus strain, the host system, time and method of harvest for antigen production, as well as on the precise conditions for carrying out the diffusion. It is also apparent that the morphological and physico-chemical properties of the antigens must be investigated to determine their relationship to structures in the infected cell and intact virion.

3. Haemagglutination by Rinderpest Virus

As already mentioned no haemagglutination (HA) has yet been demonstrated in preparations of rinderpest virus. SCOTT (1959 b) reported that KUTTLER had failed to show HA under unspecified conditions; HUYGELEN (1960 b) failed using culture-propagated, caprinised and lapinised strains and erythrocytes derived from 10 species. LIESS (1964) used virus propagated in cattle tissues and 6 isolates grown in cell cultures; he employed erythrocytes from 13 different donor species and virus fractions were prepared separately from the cells and medium of infected cultures. LIESS (1964) showed that clearcut HA occurred with infected, not normal, cell culture preparations, particularly employing rodent (rabbit and guinea pig) erythrocytes, which gave titres of up to 1/64 with the RBOK strain.

Haemagglutinating titres were highest after 4—18 hours at room temperature, using 0.5% erythrocytes; the best diluent was 0.15M phosphate buffer, pH 7.2. Treatment with Tween 80 and ether sometimes increased the titre slightly and centrifugation at 80,000 g for 2 hours lowered HA titres considerably but activity was not concentrated in the deposit. The specificity of the HA demonstrated by LIESS (1964) was unfortunately not established, e.g. by inhibition with rinderpest antisera.

At the Farcha laboratory attempts to demonstrate HA with culture-propagated (BK or BHK-21) virus of 2 strains, failed completely even after 20-fold concentration and treatment with Tween 80: ether (ANON, 1966 b). It was found, however, that pretreatment of monkey *(patas)* erythrocytes for 1 hour at 37°C with rinderpest preparations caused a loss of reactivity with measles haemagglutinin and this finding stresses the need for further studies to determine whether and in what conditions a haemagglutinin can be produced by rinderpest virus.

Haemadsorption could not be demonstrated in HeLa or chick embryo cultures infected with rinderpest virus (LIESS, 1964; ANON, 1966 b).

D. Resistance to Physical and Chemical Agents

There is a voluminous and often contradictory older literature on the stability of rinderpest virus (CURASSON, 1932; 1942), but the vast majority of the older work was only qualitative, and carried out with crude materials. It arose as an incidental requirement for the production of live virus or inactivated vaccines and as a pre-requisite for the proper decontamination of animal products and infected premises.

1. Light

THEILER (1897 b) showed that a thin layer of blood was rendered non-infectious within 2 hours of exposure to sunlight. SCOTT found that water-clear suspensions of rinderpest-infected tissues were inactivated rapidly and exponentially when placed 10 cms from a source of ultraviolet light (MACOWAN, 1956). Diluted preparations of cultured rinderpest virus were inactivated much more rapidly at 25°C or 37°C if exposed to diffused daylight or artificial lighting rather than kept in the dark (PLOWRIGHT and HERNIMAN — to be published).

2. Ultrasonic Waves

The half-life of cell culture virus of the RBOK strain was 7.5 minutes in a bath providing energy at 41.23 Kc/sec. (PLOWRIGHT, 1962a).

3. Heat

SCOTT (1959d) studied the survival of the RBOK, goat-adapted (KAG) and lapinised (NAKAMURA III) virus strains in three tissues of infected cattle which were known to maintain a reasonable pH stability (HENDERSON and BROOKSBY, 1948). Four storage temperatures were used, viz: 7°, 25°, 37° and 56°C; inactivation proceeded as a first-order reaction and the half-life periods were calculated as shown in Table 1. The rate of inactivation in heparinised blood at 37°C and 25°C was significantly slower than in other tissue suspensions. SCOTT (1964) suggested that this may have been due to the presence of high concentrations of plasma proteins or to the protection afforded by leukocytes, with which the virus is very closely associated (TODD and WHITE, 1914; DAUB-NEY, 1928). Differences between strains were not significant. The activation energy of heat inactivation was 24 kilocalories per mole, the entropy of activation being +82 calories per degree per mole.

The stability of cultured virus of the RBOK strain was investigated by PLOWRIGHT and FERRIS (1961c) and the results are given in Table 1. Stability was of the same order as for virus in cattle spleen, except at 4—7°C, and, incidentally, marginally greater than for culture-propagated measles and canine distemper viruses (BUSSELL and KARZON, 1962). The amount of protein in the suspending medium is probably of importance; thus, culture virus in maintenance medium which had been diluted 10-fold in Michaelis buffer at pH 7.2, had a half-life of 3.68 days only at 4°C (LIESS and PLOWRIGHT, 1963a) compared with a mean 9.2 days for undiluted virus (Table 1). When the serum concentration was raised to 42.5%, as in the single experiment of JOHNSON (1962c), the half-life increased to 11.5 days at 4°C (Table 1).

It is interesting to note that whereas rinderpest virus is usually stated to be "fragile" (SCOTT, 1964), this is hardly consistent with the capacity of a small fraction of tissue culture virus to survive heating at 56°C for 50—60 minutes (PLOWRIGHT and FERRIS, 1961) or exposure at 60°C for 30 minutes (DE BOER and BARBER, 1964).

4. Freezing and Storage at Low Temperature

SCOTT (1959d) found that the mean titre of blood samples supercooled to −15°C was significantly greater than that of the same samples frozen solid at this temperature; i.e. freezing and thawing caused either inactivation or aggregation of infectious virus. The slow freezing of culture virus at −25°C or −70°C in maintenance medium usually results in a considerable loss of infectivity, about 0.4 to 1.0 $\log_{10}$ units (PLOWRIGHT and FERRIS, 1961). It has been reported that these freezing losses can be reduced, as with measles virus, by the addition of 2% dimethyl sulphoxide (GRIEFF et al., 1964; ANON, 1966a).

Stocks of culture virus which had been frozen at −25°C and −70°C showed no further fall of titre over periods of 23 and 16—17 weeks, respectively, these being

Table 1. Half-life Periods of Rinderpest Virus at Different Storage Temperatures

Storage medium	Storage temperature					Reference
	60°C	56°C	37°C	25°C	4–7°C	
Cattle spleen or lymph node	—	5 minutes	105 minutes	6.4 hours	2.3 days	SCOTT (1959d)
Cattle blood	—	5 minutes	21 hours	36 hours	2.3 days	SCOTT (1959d)
LAYE/OS5[1]	—	3.5 minutes	165 minutes	—	9.2 days	PLOWRIGHT and FERRIS (1961c)
LAYH/OS10[2]	3.5 minutes	—	—	—	—	DE BOER and BARBER (1964)
LAYE — modified + vaccine stabiliser (50:50)[3]	—	—	—	16 hours	11.5 days	JOHNSON (1962c)

[1] LAYE/OS5 = 0.5% lactalbumin hydrolysate and 0.1% yeast extract in Earle's saline plus 5% ox serum.
[2] LAYH/OS10 = 0.5% lactalbumin hydrolysate and 0.1% yeast extract in Hanks' saline plus 10% ox serum.
[3] This virus contained final concentrations of constituents as follows: Lactose 3.5%; lactalbumin hydrolysate 1.5%; ox serum 42.5%. Half-life periods as calculated by SCOTT (1964).

Table 2. Half-life Periods of Cultured Rinderpest Virus, Strain RBOK, after Lyophilisation with Various Additives

Additive[1]	Storage temperature				References
	−20° to −25°C	+4°C	20° to 25°C	37°C	
Lactose-serum lactalbumin[2]	c. 6 months	c. 8 weeks	c. 1 week	—	JOHNSON (1962c)
5% lactalbumin hydrolysate	4.5 months	—	—	4.3 days	PLOWRIGHT (1963b)
2.5% lactalbumin hydrolysate	—[3]	8.9 weeks	3.6 weeks	—	PLOWRIGHT (1963b)
Nil	4 months	—	—	2.6 days	PLOWRIGHT (1963b)
?	8.5 months	10 weeks	1 week	—	ANON (1966a)

[1] All concentrations are final ones in virus-additive mixtures.
[2] See footnote to Table 1.
[3] No detectable decline in titre over periods up to 5 years (PLOWRIGHT et al. – to be published).

the maximum periods tested (PLOWRIGHT and FERRIS, 1961). A half-life period of 8.5 months has been quoted for an unspecified strain of virus in culture fluids at $-22°C$ (ANON, 1966a).

5. Lyophilisation

Virus which had been rapidly frozen and dried from the frozen state has a far greater stability than virus in liquid media (JACOTOT, 1932a). The majority of published data refer to vaccine virus and, unfortunately, seldom allow the calculation of an inactivation rate. SCOTT (1964) used the figures of NGUYEN-BA-LUONG et al. (1958) for freeze-dried lapinised virus to calculate half-life periods of 0.8 to 1.1 days at 37°C, 2 days at 4°C and 23—38 days at $-22°C$; with the exception of 37°C, therefore, stability was lower than in wet tissues or in liquid media, a result which should be treated with scepticism.

Published data for freeze-dried cell-culture virus are given in Table 2, but continued observations on numerous vaccine batches stored at $-20°C$ over a 4—6 year period failed to reveal any significant decline of titre, either without or with the use of additives; in addition virus lyophilised with a lactalbumin-sucrose additive was stable at 4°C over a period of about 2.5 years (PLOWRIGHT, TAYLOR and RAMPTON — to be published). The discrepancies between these older and more recent findings are probably due to uncontrolled variations in the sensitivity of the cell cultures employed for titrations; there is also a large variance in titres estimated by using 10-fold dilutions of virus — 0.28 to 0.50 $\log_{10}$ units, occasionally even higher. Finally a close control is necessary on the residual moisture content and the degree of vacuum in ampoules of freeze-dried virus if data of diverse origin are to be compared (SCOTT, 1964).

Losses in the lyophilisation of vaccines were initially 90% or more (ANON, 1966a) but could be reduced by, for example, the addition of defibrinated or citrated blood in drying lapinised rinderpest virus (CHENG and FISCHMAN, 1949) or a modified *Mist. dessicans* of FRY and GREAVES (1951) for goat-adapted virus (PROVOST et al., 1958). The mean loss of titre in drying culture-propagated virus without additives was 1.57 log units, with *Mist. dessicans* it was still 1.15 units but losses were reduced to a mean 0.58 and 0.85 units respectively by the addition of a final concentration of 0.5% or 0.25% of lactalbumin hydrolysate (PLOW-RIGHT, 1963b).

6. pH Stability

Earlier work which may occasionally give some indication of the effect of pH on the stability of rinderpest virus in infected blood or tissues has been reviewed at length by CURASSON (1932; 1942). In these experiments it was seldom or never possible to dissociate pH effects from other uncontrolled factors but it was generally agreed that putrefaction and the acidification which occurs as a result of autolysis in meat, caused rapid destruction of the virus.

The first systematic attempt to find the optimal pH for the survival of rinderpest virus was made by MAURER (1946). Using 1% tissue suspensions in SØRENSEN's m/10 phosphate buffers he found that pH 6.0 or 7.0 gave better virus survival than pH 8.0, infectivity persisting for 44 (<48—50) days in the two former but only for 10 days (<14) in the latter; he suggested that the exact optimum lays between pH 6.5—7.0. LIESS and PLOWRIGHT (1963a) used culture fluids infected with the RBOK strain after 95 culture passages and also two recent field isolates after 2—3 passages only in calf kidney cells; they diluted virus preparations 10-fold in MICHAELIS buffers at 4°C and found that high-passage virus was relatively stable between pH 4.0 to 10.2. Inactivation was exponential, the half-life at the extremes of the range being 2.3 to 2.6 hours. Stability was greatest between pH 7.2 and 7.9 with a half-life of about 3.7 days. Inactivation was very rapid at pH 3.0, the half-life for the RBOK strain being 24 secs., whereas it was significantly shorter (12.5 secs.) for the virulent RGK/1 isolate; recent isolates were also far more sensitive than the RBOK strain at pH 4.0 and 5.0 but not at pH 10.1 and 10.7.

DE BOER and BARBER (1964) working with the Pendik isolate at 26°C, in culture medium containing 10% ox serum, found that all infectivity disappeared within one and a half to two minutes at pH 2.0 and within 10 minutes at pH 3.0 or 12.0; rates of inactivation were not determined. At pH 3.5 and 4.0 or 10.0 and 11.0 the data provided by DE BOER and BARBER (1964) show a definite heterogeneity of the virus population but their regression equations do not appear to take this into account, neither do they comment on this feature. It is unfortunate that their figures have already been used in calculating the half-life of surviving virus at 26°C (ANON, 1966a).

7. Salt Concentration

Earlier work on the effect of high concentrations of sodium chloride on the virus in meat or hides has been reviewed by CURASSON (1932; 1942) — in general, infectivity disappeared rapidly. According to MAURER (1946) the optimal molarity for 1% cattle spleen suspensions in phosphate buffers of pH 7.0 was 0.1 M; buffers of 1.0 or 0.01 M were definitely inferior. It has recently been demonstrated (PLOWRIGHT and HERNIMAN, 1967) that culture virus of the RBOK strain loses about 66% of its infectivity immediately after 200-fold dilution in water at 4°C. This loss does not occur when a comparable dilution is prepared in 0.85% sodium chloride, but is increased to 85% and 92% if the water is at 25°C or 37°C respectively; it occurs irrespectively of whether previously lyophilised or freshly-harvested virus is used. A smaller loss of infectivity occurs when virus is diluted 20-fold in water. This very rapid loss of infectivity suggests osmotic disruption of a large proportion of infective virus particles.

The effect on rinderpest virus of high concentrations of sulphate ions has been studied by several groups of workers, following original observations with measles virus (RAPP et al., 1965). When rinderpest culture vaccine was diluted to field strength (probably 100-fold) in water and 1 M magnesium sulphate at 37°C, the half-life of the infectivity was 24 minutes and 6 hours respectively (ANON, 1966a, b). On the other hand the effect of 1 M magnesium chloride was a sensitising one, similar to that on measles virus; in this solution at 50°C, virtually all virus had been inactivated within 30 minutes, compared with 90% survival in 1 M magnesium sulphate and 0.01% survival in water. Sodium sulphate (1 M) had a protective action similar to that of magnesium sulphate and the effect of both salts declined with concentrations below 1.0 M (ROBIN and BOURDIN, 1966). The rate of inactivation of virus diluted 2-fold in water at 50°C was 3.5 times greater than that in a concentration of 1 M $MgSO_4$, the respective half-lives being 21 minutes and 74 minutes (PLOWRIGHT and HERNIMAN, 1967).

8. Glycerol

As noted by SCOTT (1964) the sensitivity of rinderpest virus to glycerol was first commented on by SEMMER in 1896 und confirmed by ROBERT KOCH in the following year; it was also utilised in the preparation of the first inactivated tissue vaccine (KAKIZAKI, 1918). No observations appear to have been recorded on the glycerol-sensitivity of measles or distemper viruses.

9. Lipid Solvents

Rinderpest infectivity, as would be expected of a myxovirus, is completely inactivated by treatment overnight at 4°C with 20% by volume of ethyl ether and also by 5% chloroform in 10 minutes at 22°C (PLOWRIGHT, 1962a). Chloroform was widely used in the past for the preparations of inactivated tissue vaccines (KELSER et al., 1929; WALKER et al., 1946b) and toluol was used for the same purpose, sometimes mixed with glycerol (CURASSON, 1942).

The use of bile in the preparation of killed vaccines (CURASSON, 1942) may depend on a similar effect of bile salts on the lipids of the viral membrane.

10. Disinfectants

Phenol, chinosol and formalin have been very widely used to render rinderpest-infected tissues non-infective without loss of antigenicity (JACOTOT, 1950). No work has been published on quantitative aspects of the inactivation process.

11. β-propiolactone

STONE and DeLAY (1961) showed that suspensions of cattle or rabbit spleens containing $10^{4.7}$ lethal doses of rinderpest virus per ml were rendered non-infectious within 30 minutes by treatment with 0.4% β-propiolactone (BPL) at 28°C; normal bovine serum (50% v/v) did not prevent the inactivation. More recently it has been found that exposure to 0.1% BPL for 18 hours at 4°C also completely inactivates the infectivity without affecting the immunogenicity (ANON, 1966a).

12. Hydroxylamine

Treatment with 1M hydroxylamine for 15 minutes at 22°C and pH 7.0 (FRANKLIN and WECKER, 1959) resulted in a loss of over 90% of the infectivity of high culture-passage virus of the RBOK strain (PLOWRIGHT, 1962a). Following the demonstration that the Beaudette strain of NDV, of low virulence, was susceptible to this agent (ROTT and SCHÄFER, 1962) whereas a highly pathogenic strain, 'Italien', was not (SCHÄFER and ROTT, 1962), the rate of inactivation by 1M hydroxylamine of the attenuated variant of RBOK rinderpest virus was compared with that of the virulent RGK/1 strain (LIESS and PLOWRIGHT, 1964). Loss of infectivity proceeded as a first-order reaction, the attenuated strain being inactivated about 3 times as rapidly as the virulent isolate; the half-lives were 26 and 74 minutes respectively. This observation needs to be extended to further strains and correlated, as for strains of NDV (ROTT and SCHÄFER, 1962) with the site of intracellular virus maturation, which may influence susceptibility to hydroxylamine. It has already been shown that two virulent strains of rinderpest virus (low-passage RBOK and RGK/1), tend to accumulate within calf kidney cells, whereas the attenuated RBOK variant is released more rapidly (PLOW-RIGHT, 1964b).

13. Trypsin

Two strains of rinderpest virus, RBOK and RGK/1, were inactivated within 15 minutes at 37°C by 0.5 mg/ml of crystalline trypsin in a serum-free medium; the RBOK strain was found to include a small resistant fraction, equal to 0.01% of the original infectivity (PLOWRIGHT, 1964b). In its trypsin sensitivity rinderpest virus resembles measles and Sendai viruses; the effect of trypsin is possibly to split a protein in the viral envelope, and in this context it is interesting to note that measles haemagglutinin is also rapidly destroyed by trypsin (NORRBY, 1962b).

E. Cultivation

1. History

It is remarkable that rinderpest virus was probably first cultivated *in vitro* over 50 years ago (BOYNTON, 1914) at about the same time as the first successful attempts were made to cultivate vaccinia and poliomyelitis in plasma clot preparations of surviving tissue (SANDERS et al., 1953). BOYNTON (1914) used 10 ml quantities of defibrinated normal ox blood, with the addition of 0.3% glucose. The surface was sealed with sterile paraffin and the inoculum consisted of 0.5 ml of blood from cattle. Incubation was at 40°C and sub-inoculation was carried out at intervals of 3—4 days; in each of 2 series 6 transfers were successfully effected, infectivity for cattle being demonstrated after 19—21 days, involving estimated dilution of the original inocula, by about 28×10^6 times. This experiment undoubtedly implied proliferation of the virus but could not be repeated successfully by MINETT (1923) or DAUBNEY (1928). The latter author also investigated the suspended-tissue technique, employing rinderpest-infected spleen fragments in serum broth, but failed to demonstrate persistence of virus. DAUBNEY (1938) claimed limited success with Rivers-type cultures of embryonic and post-natal tissues of cattle, goats and rabbits but CARMICHAEL (1939) failed with fragments of calf spleen and testis or chick embryo spleen.

TAKEMATSU and MORIMOTO (1954) reported that lapinised rinderpest virus proliferated in roller-tube cultures of rabbit spleen, lymph node and bone marrow but no cytopathic effects were reported and NAKAMURA et al. (1958) showed that the lapinised-avianised strain of NAKAMURA and MIYAMOTO (1953) multiplied in suspended-fragment cultures of chick embryos, but it was necessary to titrate the virus by sub-inoculation into eggs, the embryonic spleens of which were then used for complement-fixation tests. Nevertheless, the system was applied to the detection and titration of rinderpest-neutralising antibody (NAKAMURA, 1957b). Other attempts to demonstrate the growth of goat-adapted and lapinised strains of virus in suspended-fragment and plasma-clot cultures of goat or rabbit tissues were not successful (PIERCY and FERRIS, 1957). The proliferation of rinderpest virus in monolayer tissue cultures with the production of a cytopathic effect was first reported in 1957 (PLOWRIGHT and FERRIS, 1957).

2. Host-cell Range

The range of monolayer cell cultures which support the proliferation of bovine strains of rinderpest virus are shown in Table 3. It is evident that the virus grows readily in monolayers of epithelial or fibroblastic cells from many

Table 3. *The Host-cell Range of Bovine Strains of Rinderpest Virus in Primary Monolayer Cultures*

Cell type	Virus growth	Cytopathic effects	Reference
Bovine			
Embryonic kidney	+	+	PLOWRIGHT and FERRIS (1959a)
Post-natal kidney	+	+	PLOWRIGHT and FERRIS (1959a)
Embryonic skin-muscle	+	+	PLOWRIGHT and FERRIS (1959a)
Testis	+	+	PLOWRIGHT and FERRIS (1959a)
Thyroid	+	+	PLOWRIGHT and FERRIS (1961a)
Leukocytes	+	+[1]	TOKUDA et al. (1962; 1963)
Sheep			
Embryonic kidney	+	+	PLOWRIGHT and FERRIS (1959a)
Testis	+	+	PLOWRIGHT and FERRIS (1959a)
Goat			
Kidney	+	+	PLOWRIGHT and FERRIS (1959a)
Pig			
Embryonic kidney	+	+	PLOWRIGHT and FERRIS (1959a)
Post-natal kidney	+	+	MALMQUIST (1959)
Dog			
Kidney	+	+	PLOWRIGHT (1962a); SINGH and EL CICY (1966)
Rabbit			
Embryonic kidney	−	−	PLOWRIGHT and FERRIS (1959a)
Testis	−	−	PLOWRIGHT and FERRIS (1959a)
Post-natal kidney	−	−	ISOGAI (1961)
Hamster			
Kidney	+	+	PLOWRIGHT and FERRIS (1959a)
Monkey (C. aethiops)			
Kidney	+	+	PLOWRIGHT (unpublished)
Human			
Amnion	+	−	PLOWRIGHT (1962)
Chick Embryo			
Skin-muscle	+	+	ISOGAI (1961)
Kidney	?	+	ISOGAI (1961)

[1] Detected in stained preparations only.

susceptible species, including ruminants, swine, dog, hamster and chick embryos. Primary kidney cultures from *Cercopithecus aethiops* were highly susceptible to two virulent strains, RBOK and RGK/1 (PLOWRIGHT — unpublished) and infectivity persisted for 42 and 66 days in human amnion cultures, without the production of gross cytopathic effects (PLOWRIGHT, 1962a); man has not yet been tested for susceptibility to rinderpest virus, although this could clearly be of some interest. The most surprising feature was the failure of the bovine virus to grow in kidney, embryonic kidney or testis monolayers derived from rabbits (PLOWRIGHT and FERRIS, 1959a; HUYGELEN, 1960a; ISOGAI, 1961), in spite of the well-known susceptibility of this species, at least to some virulent strains of virus (vide infra). The most susceptible cells are undoubtedly those derived from bovine kidney and not leukocytes as has been suggested (ANON, 1966a); data provided by TOKUDA et al. (1963) show a difference in sensitivity ranging from 2 to 10-fold in favour of kidney cells.

Table 4. *Serially Cultivated Cells Susceptible to Rinderpest Virus*

Cell origin	Cell reference	Virus growth	Cytopathic effects	Reference
Calf kidney	—	+	+	PLOWRIGHT and FERRIS (1961b)
Calf kidney	—	+	+	JOHNSON (1962a)
Calf kidney	MADIN and DARBY (1958)	+	+	JOHNSON (1962a)
Calf kidney	BRION and GRUEST, 1957	+	+	PLOWRIGHT (1962a)
Pig kidney	MADIN (1959)	+	+	PLOWRIGHT (1962a)
Hamster kidney	BHK 21 — C 13	+	+	ANON (1966a)
Dog kidney	MADIN (1959)	+	+	PLOWRIGHT (1962a)
Rabbit kidney	DREW (1957)	+	+	PLOWRIGHT (1962a)
Human Neoplastic	HeLa	+	+	LIESS and PLOWRIGHT (1963a)
Human Neoplastic	KB	+	+	ANON (1966a)
Monkey kidney	MS	+	+	SINGH and EL CICY (1966)

Numerous serially-cultivated cells, the majority of them established "cell-lines" (HAYFLICK and MOORHEAD, 1961) have been used for the production of rinderpest virus (Table 4). Many strains of serially-cultivated calf kidney cells have been found useful in serological work or in the production of vaccine virus (PLOWRIGHT and FERRIS, 1961b; JOHNSON, 1962a, b).

Japanese workers, particularly, have investigated the susceptibility of various monolayer cell cultures to attenuated strains of rinderpest; their results are given, together with those of earlier studies, in Table 5. The difficulty of adapting the older, goat-adapted and lapinised strains to growth in many types of cell culture, with the possible exception of leukocytes, may be a useful marker.

3. Cytopathogenicity

a) General

The cytopathic effects of rinderpest virus in monolayers of calf kidney (BK) cells are most extensive and striking in areas of the cell sheet which are extending and consist of "younger" and dividing cells. Hence, in roller-tube cultures they almost invariably begin towards the edge of the expanding cell sheet so long as an adequate growth medium is provided; again, cytopathic effects become more extensive and, incidentally, virus yields are higher if virus is seeded with trypsin-dispersed cells into the culture vessels rather than inoculated into already completed monolayers. Cytopathic effects tend to be more progressive and virus yields are higher when the vessels are rolled than if they are incubated stationary (PLOW-RIGHT and FERRIS, 1957; 1959a; — unpublished observations).

Table 5. *The Host-cell Range of Some Attenuated Vaccine Strains of Rinderpest Virus (after* NAKAMURA, 1965*)*

Author	Cell type	Caprinised (KAG)	Lapinised (Nakamura III)	Lapinised Avianised (LA)	Avianised (BA)	Lapinised BEK-adapted
PLOWRIGHT and FERRIS (1959a, 1962a)	Goat, bovine kidney	--	—	NT	NT	NT
	Rabbit, kidney testis	--	—	NT	NT	NT
ISOGAI (1961)	Bovine kidney	NT	±[1]	+	+	+[1]
	Bovine testis	NT	±	+	+	NT
	Rabbit kidney	NT	—	—	—	NT
	Chick embryo.	NT	—	+	+	NT
	Chick embryonic kidney	NT	—	+	+	NT
TOKUDA et al. (1962b)	Bovine leukocyte	NT	+	—	NT	NT
NAKAMURA (1965)	Dog kidney	NT	+	+	NT	+

[1] Positive in embryonic kidney only. NT = not tested.

Using the RBOK strain of virus as inoculum, cytopathic effects were not at first detected in living preparations during the first 4 passages in stationary cultures, although they were demonstrable in stained preparations by the 3rd day of the first passage. When cells were infected as trypsin-dispersed suspensions, characteristic cytopathic changes were seen as early as the 3rd day (PLOW-RIGHT and FERRIS, 1959a). It was later shown, however, that cattle tissues infected with the same strain of virus caused typical cytopathic effects in established roller-tube cultures of BK cells, detectable by the 3rd to 6th days post-inoculation. These effects progressed rapidly to involve 50—80% of the cell sheet by the 8th to 9th days but minimal quantities of virus did not produce detectable cell changes until the 11th or 12th days; virus titrations could be performed directly in BK cultures (PLOWRIGHT and FERRIS, 1962a).

DE BOER (1960; 1961) "adapted" the Pak Chong (Siamese buffalo) and

Pendik (Turkish) laboratory strains to growth in BK cells; bovine spleen suspensions were mixed with trypsin-dispersed cells at the same time as dispensing the latter into culture vessels but cytopathic changes did not become visible until the 18th to 21st days; passages were made in a similar manner, the time to appearance of cytopathic effects decreasing to 7 days. In titrating virus DE BOER (1961) did not obtain end-points until after 21 days but this, together with the slow "adaptation" and very long time to peak yields, may well have been due to the use of stationary cultures. PROVOST and VILLEMOT (1961) and GILBERT and MONNIER (1962a) in West Africa used locally-maintained, virulent strains of virus (Farcha and Dakar, respectively) which they established readily in bovine embryonic kidney (BEK) cells, inoculated either as suspensions or as partially-grown monolayers; cytopathic effects were seen after 2 or 7 days, respectively. In East Africa, four strains of virus were isolated in BK cultures from cattle in 1961 (PLOWRIGHT and FERRIS, 1962a; PLOWRIGHT, 1963a), another one from a buffalo *(Syncerus caffer)*, one from a bushbuck *(Tragelaphus scriptus)* (LIESS, 1963), and one (RGK/1) from a reticulated giraffe *(Giraffa reticulata)* (PLOWRIGHT, 1963c). Several further isolations have since been made from cattle (PLOWRIGHT; TAYLOR; — unpublished) and it seems that all virulent field strains can readily be detected and titrated in BK cultures; serial passage presents no difficulties. The KAG (caprinised) strain did not, however, cause cytopathic effects in roller-tube cultures of BK cells (PLOWRIGHT and FERRIS, 1962a) and this may be a useful *in vitro* method of differentiating it from virulent strains. GILBERT and MONNIER (1962b) cultivated one strain of the immunologically identical virus of "peste des petits ruminants" in sheep embryonic kidney cells, although they failed in several other attempts.

b) Cytopathology

The cytopathogenic effect of rinderpest virus is characterised by the formation of multinucleated giant cells (syncytia) and by the appearance of stellate or spindle cells with long, fine, often anastomosing processes (Fig. 1); infected cells frequently contained eosinophilic cytoplasmic and intranuclear inclusions. These different aspects of the cytopathology will be dealt with separately:

(i) Syncytia

Depending on the virus strain, the cell type and the stage of infection these may vary from small, angular or rounded, refractile structures containing 2 to 3 or more nuclei (Figs. 2 and 3) to large sheets of cells containing several hundred nuclei and with ill-defined edges to the common cytoplasm. In BK cultures infected at seeding with large inocula of the RBOK strain of virus, there are often two cycles of syncytium formation. The first, beginning on the 4th or 5th days and reaching its maximum on the 6th to 10th days, is characterised by large numbers of relatively small, heavily-vacuolated syncytia containing a central, granular mass including nuclei. The second cycle, reaching its peak about the 14th to 18th days, consists largely of the fusion of small, little-vacuolated syncytia and single cells; these expand and form extensive, glassy sheets containing one or more centrally-located clumps or rings of nuclei with large, homogenous cytoplasmic inclusions surrounding them (Fig. 4). The centre of the nuclear ring is often occupied by amorphous, granular material.

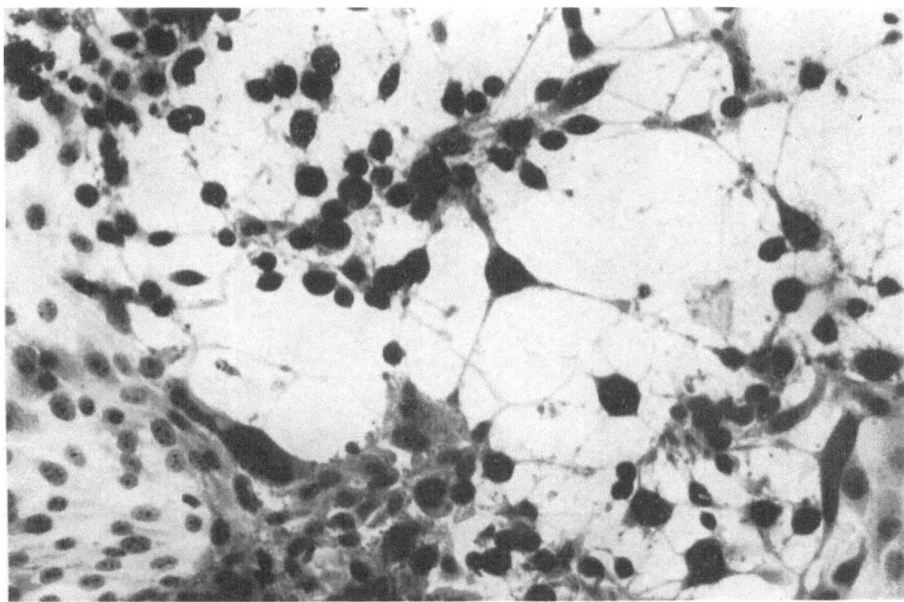

Fig. 1. Monolayer of calf kidney (BK) cells infected 7 days previously with the RBOK strain of rinderpest virus. Early type of cytopathic effect with small, rounded and stellate elements, often joined by long, fibrillary processes. Bouin: H and E (× 350).

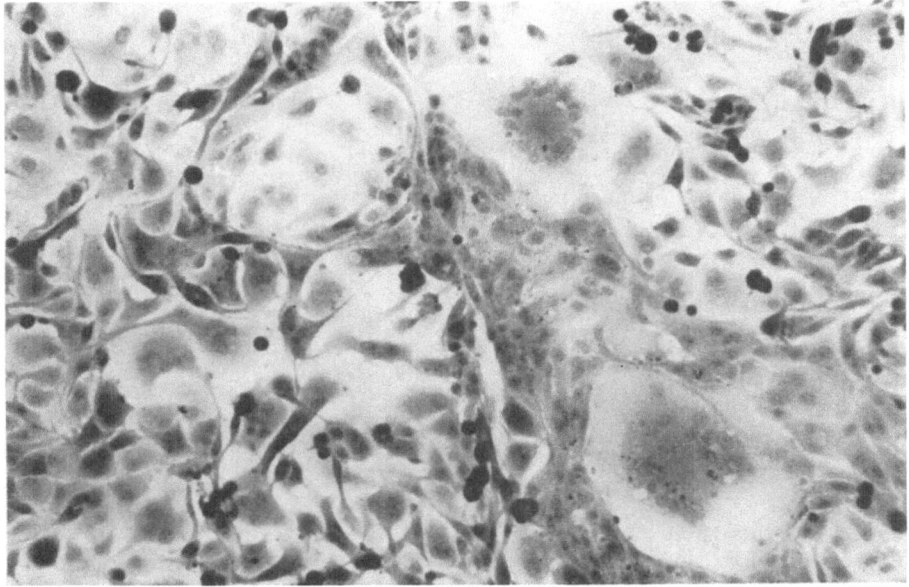

Fig. 2. BK monolayer 9 days after infection with a buffalo strain of virus. Large rounded syncytia with a central ring of nuclei are mixed with denser, small, angular syncytia and rounded cells. Bouin: H and E (× 200).

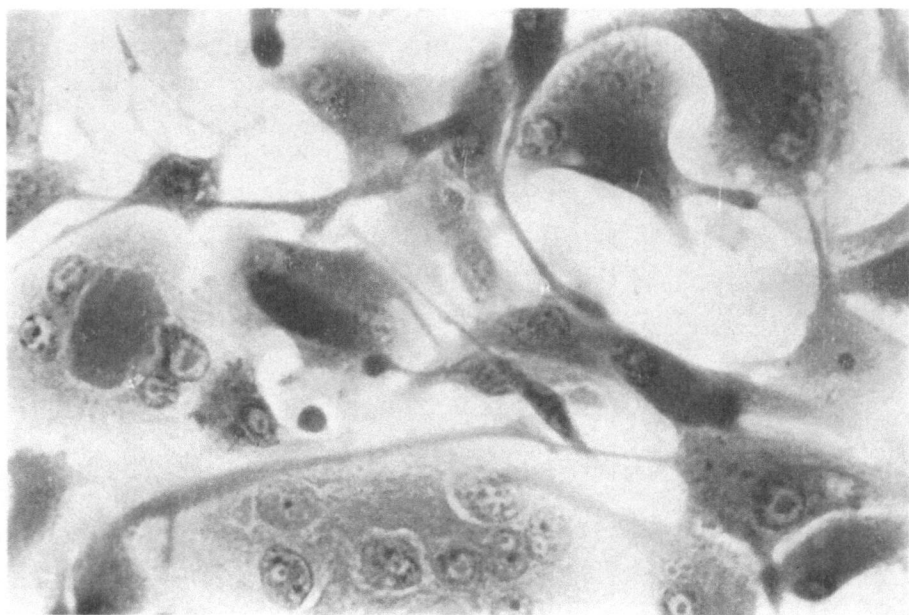

Fig. 3. Same culture as Fig. 2 but 14 days after infection. Stellate syncytia with anastomosing processes; the one on the left shows a large cytoplasmic inclusion and small intranuclear inclusions are abundant. Bouin: H and E (×700).

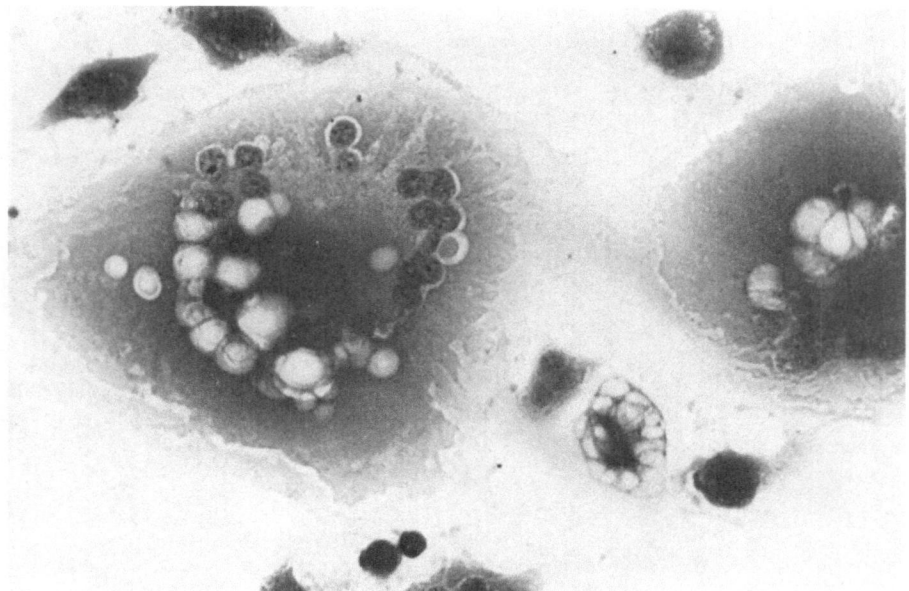

Fig. 4. Large syncytia of the second phase, 14 days after infection of a BK monolayer with the RBOK strain. Note the ring of nuclei surrounded by very extensive sheets of cytoplasmic inclusion material. The two large syncytia are beginning to fuse together. Bouin: H and E (×280).

The first cycle of syncytium formation is accompanied by the release of virus to high titre (c. $10^{5.5}$ to $10^{6.5}$ per ml); the second by yields 100-fold or more lower (PLOWRIGHT and FERRIS, 1959a; JOHNSON, 1962a). A similar sequence of cytopathic changes has been described by GILBERT and MONNIER (1962a) with the Dakar strain growing at 37°C, whereas larger syncytia formed at 40°C. The cytopathic effects of the virus of "peste des petits ruminants" in sheep embryonic kidney cells at 40°C was also dominated by the formation of "enormous giant cells" (GILBERT and MONNIER, 1962b). According to PROVOST and VILLEMOT (1961) the smaller vacuolated type of syncytium sometimes contained nuclei in an abnormal type of mitosis but the larger type with a ring of nuclei did not and must have been derived, therefore, from the fusion of previously-formed single cells. It is difficult to accept this suggestion without more convincing evidence, but the more complete observations of ROIZMAN and SCHLUEDERBERG (1962), on measles virus growing in HEp-2 cells, do provide support for the conception that infection of mitotic cells by these viruses may give rise to the smaller, stellate type of giant cell, incapable of "recruiting" further cells into itself. The laboratory segregation of virus lines causing predominantly "strand formation", i.e. the production of small stellate or spindle-shaped syncytia has not been described for rinderpest as it has for measles virus (e.g. ODDO et al., 1961), but it has been found that small "stellate" syncytia are often produced in profusion in BK cultures infected with recent field isolates of low cattle virulence (Figs. 2 and 3) (PLOWRIGHT, 1962a; 1963a). The strain RTK/1, of bushbuck origin, was particularly striking in this respect (LIESS, 1963).

The omission of glutamine from the maintenance medium of BK cultures which were used for the growth of rinderpest virus did not cause the production of large syncytia to dominate the cytopathic effect, as was reported for measles virus in HEp-2 cells (REISSIG et al., 1956; PLOWRIGHT and FERRIS, 1959a). On the other hand, environmental factors may play a part in determining the predominant type of cytopathic change, since LIESS and PLOWRIGHT (1963b) found that the substitution of 10% lamb serum for 10% ox serum, in the growth and maintenance medium of HeLa cells, immediately increased the numbers of large syncytia but may have depressed virus yield, instead of increasing it as with measles virus (MOURA, 1962).

The syncytia forming in rinderpest-infected cultures of bovine leukocytes were stated to be larger and more "amorphous" than those which developed in uninfected cultures (TOKUDA et al., 1962).

(ii) Cytoplasmic Inclusions

Masses of a deeply eosinophilic material were seen in 1957 in the cytoplasm of rinderpest-induced syncytia, whether in BK or fibroblastic cell cultures; a gradual increase in these aggregations in ageing syncytia was noted in 1959 (Fig. 4) (PLOWRIGHT and FERRIS, 1957; 1959a). The earliest inclusions in BK cells were small, granular and outlined by a narrow clear zone, but later they enlarged, fused and became more homogenous, forming continuous masses surrounding the nuclei of syncytia but still usually with well-defined edges. Similar masses were depicted or described by HUYGELEN (1960a) in calf testis cells and by GILBERT and MONNIER (1962a, b) in kidney cells, while TOKUDA et al. (1962) found them in leukocyte cultures infected with virulent, lapinised, LA or BA strains of virus. LIESS and PLOWRIGHT (1963 b) saw clearly demarcated inclusions, at first weakly basophilic but later neutrophilic in infected HeLa cells fixed in Carnoy's fluid.

Cytochemical and immunofluorescence studies on BK and HeLa cells (LIESS, 1964) showed clearly that the early basophilia of the inclusions in preparations stained by the May-Grünwald:Giemsa technique (JACOBSON and WEBB, 1952) could be removed differentially and completely by RNase treatment, but that an eosinophilic matrix still remained; the early inclusions were also sites of virus antigen and, incidentally, could be seen in unfixed HeLa cells. The inclusions contained no Feulgen-positive material and stained intensely with pyronin in Brachet's test; in preparations stained with acridine orange they fluoresced intensely red (LIESS, 1964; PROVOST et al., 1965 b).

BREESE and DE BOER (1963) studied ultrathin sections of BK cells infected at high multiplicity with the virulent Pendik strain of virus; they claimed to have seen the first indications of virus-induced changes at 3 hours after infection, in the form of irregular osmiophilic bodies in the cytoplasm and even small syncytia; at 24 hours and later they saw indications of viral development within mitochondria and two types of osmiophilic particles, of diameters 400—600 and 1500—3000 Å, respectively; the latter were considered to be rinderpest virions. In direct contrast, PROVOST et al. (1965 b) could find no changes in the mitochondria of cells infected with the virulent Farcha strain of virus; they considered that the cytoplasmic "inclusions" seen in the light microscope were composite structures formed from hypertrophied ergastoplasmic vesicles, mitochondria and free nucleocapsids. Ribosomes accumulated near the ergastoplasmic vesicles, in which electron-opaque particles of 500—750 Å were formed from about the 6th hour post-infection, being sometimes freed on rupture of the vesicles. Virions were released from the cell surface through the agency of microvilli, which contributed plasma membrane to the viral envelope. BREESE and DE BOER (1963) did not observe particles leaving the cell but suggested escape following its disruption.

TAJIMA et al. (1967), using ferritin-tagged antibody and negative contrast electronmicroscopy, examined both infected bovine kidney cells in cultures and also mucosal and lymphoid tissues of infected cattle and rabbits. The cytoplasmic inclusions in monolayers and rabbit tissues consisted of filaments arranged randomly or roughly parallel with a diameter ca. 200 Å and a clear central canal about 70 Å wide. A second type of inclusion, found most commonly in cattle tissues, consisted of compact aggregates of fine fibrillary material often so densely packed that it was difficult to see the filaments; in cultured cells inclusions contained filaments up to 4000 Å long, of 190 Å diameter, with a central canal ca. 50 Å wide and external serrations with a periodicity ca. 65 Å. These filaments showed an obvious close resemblance to the internal component of rinderpest and related viruses, whilst ferritin-tagged antibody accumulated within and near the periphery of the inclusions. TAJIMA et al. (1967) saw no circular structures comparable to intact virions, either within or at the surface of cells.

It is difficult to reconcile the observations of BREESE and DE BOER (1963) with those of PROVOST et al. (1965 b) or with the undoubted affinities of rinderpest to other large myxoviruses. The structures described by TAJIMA et al. (1967)

differ from those seen by the other groups and more work is evidently necessary to clarify the intracellular developmental cycle of rinderpest virus.

(iii) Intranuclear Inclusions

Small, rounded eosinophilic bodies delineated by a partially-cleared halo, were found in the nuclei of some syncytia in the later stages of infection with the RBOK strain of virus (PLOWRIGHT and FERRIS, 1957; 1959a). Single nuclei usually contained one or two of these bodies but sometimes four or five were detectable; the larger ones sometimes showed an irregularity of structure, suggesting vacuolation but the smaller ones were homogenous. Margination of chromatin and displacement of nucleoli did not usually occur with this strain of virus and the inclusions were therefore described as of type B (COWDRY, 1934). GILBERT and MONNIER (1962a), using the Dakar strain of virus growing in BK cells at 37°C found that intranuclear inclusions appeared on the 5th to 6th day, and were present in the majority of nuclei by the 8th day. They varied greatly in size but some occupied about two-thirds of the nuclear volume; at 40°C the inclusions were larger and more frequent. The virus of "peste des petits ruminants" produced similar intranuclear inclusions in the majority of sheep kidney cells at 37°C and they eventually occupied most of the nuclear volume; at 40°C they developed more rapidly and affected virtually all cells (GILBERT and MONNIER, 1962b). TOKUDA et al. (1962) observed intranuclear inclusions in leukocyte cultures infected with virulent and attenuated vaccine strains.

PLOWRIGHT (1963a) found that some recent field isolates, particularly a buffalo strain, produced in BK cultures intranuclear inclusions which appeared earlier and were larger and more frequent than those induced by the RBOK strain (Figs. 5 and 6). LIESS (1964), working with this buffalo strain, found that the nuclear inclusions always fluoresced a reddish colour in acridine orange preparations and otherwise stained typically as ribonucleoprotein; the intranuclear inclusions have not been identified with foci of specific immunofluorescence. No structures corresponding to the intranuclear inclusions were seen by BREESE and DE BOER (1963) or PROVOST et al. (1965b) in electronmicrographs of osmic-fixed cells.

4. The Study of Rinderpest-infected Cells by Immunofluorescence

Using the indirect, fluorescent-antibody method of WELLER and COONS (1954) and hyperimmune ox serum it was found that fluorescent granules first appeared in the cytoplasm only, between 6 and 12 hours after infection of primary BK cells (LIESS, 1963). Occasionally a network of fluorescent filaments was seen and it is tempting now to suggest that these represented channels of the endoplasmic reticulum in which early virus synthesis occurred (PROVOST et al., 1965b). The number of affected cells as well as the number and size of the granules increased considerably by 24 hours and antigen was also found in the anastomosing processes which connected some single cells, or in the cytoplasm of small syncytia. Still later, in heavily vacuolated syncytia induced by the RBOK strain, there was intense fluorescence of the compressed strands of cytoplasm between the vacuoles; in the later stages of infection by a bushbuck strain of virus (RTK/1) it was found that the plasma membrane of spindle-shaped and stellate cells showed an intense, diffuse fluorescence.

In ageing syncytia with large, cytoplasmic inclusions the latter sometimes fluoresced specifically but usually less strongly than in earlier stages of the infection; in HeLa cells it was clearly shown that the well-defined sites of cyto-plasmic fluorescence corresponded with the inclusions as seen in stained pre-parations. Incidentally, it was shown that measles antigen in the cytoplasm of HeLa cells reacted specifically with antibody in rinderpest-immune cattle serum, thus demonstrating the close relationship between these two viruses (LIESS, 1963; LIESS and PLOWRIGHT, 1963b).

At first, antigen was not found in the nuclei except for a few small granules in the later stages of infection (LIESS, 1963); re-investigation, however, using air-dried instead of acetone-fixed ($-30°$C) preparations, revealed fluorescent particles in the nucleus as early as 8 hours after infection often accompanied by perinuclear fluorescence (LIESS, 1964). Cytoplasmic fluorescence was first de-tected at 19 hours and the eclipse phase in this experiment was probably 12 hours; comparable acetone-fixed cultures showed only cytoplasmic fluorescence, from 12 hours onwards, whilst the nuclear fluorescence was not detected even after 24 hours. LIESS (1964) concluded that the first synthesis of virus-specific materials probably occurred in the nucleus.

5. Plaque Formation by Rinderpest Virus

It was reported that rinderpest virus growing in calf testis monolayers pro-duced rounded clearings of about 1 mm diameter, so long as dissemination of virus through the fluid was prevented by the inclusion of immune serum; the number of clearings was proportional to the dilution of virus (PLOWRIGHT, 1962a). The production of plaques by Pendik strain virus on monolayers of primary BK cells was described by McKERCHER (1963); he used a 30-minute adsorption period and an agar-overlay in medical flat bottles incubated at 37°C. Plaques became visible by 7—8 days and attained an average diameter of 3 mms by the 12th day; after a further 7 days the size increased to 5 mms. The optimal concentration of Noble agar was 1.0—1.5% and neutral red at 1/5000 concentration had no inhibitory effect on plaque formation; omission of serum from the overlay was also without effect. McKERCHER (1963) also found a com-parable number of plaques when virus was mixed with trypsin-dispersed cells at the time of seeding into bottles; the overlay was added on the 5th or 6th days. Virus assay by plaque formation, unfortunately gave considerably lower titres (probably about 2 $\log_{10}$ units) than 50% dilution endpoints in monolayers.

In a later study, McKERCHER (1964b) found that the optimal size of in-oculum for prescription bottles of 4 fl. ozs. capacity was 0.5 ml. The rinderpest plaques consisted at first of intact but dead cells which failed to take up neutral red and, later, of amorphous cellular debris; they had a uniform size and well-defined circular edge. Plaque inhibition by specific immune serum was used in developing a technique for virus identification.

There is an obvious need for much further investigation of rinderpest viruses by the plaque technique, especially in relation to the analysis of virus populations for clones of varying animal virulence; stability and growth studies, also, would be considerably facilitated by a sensitive plaque assay.

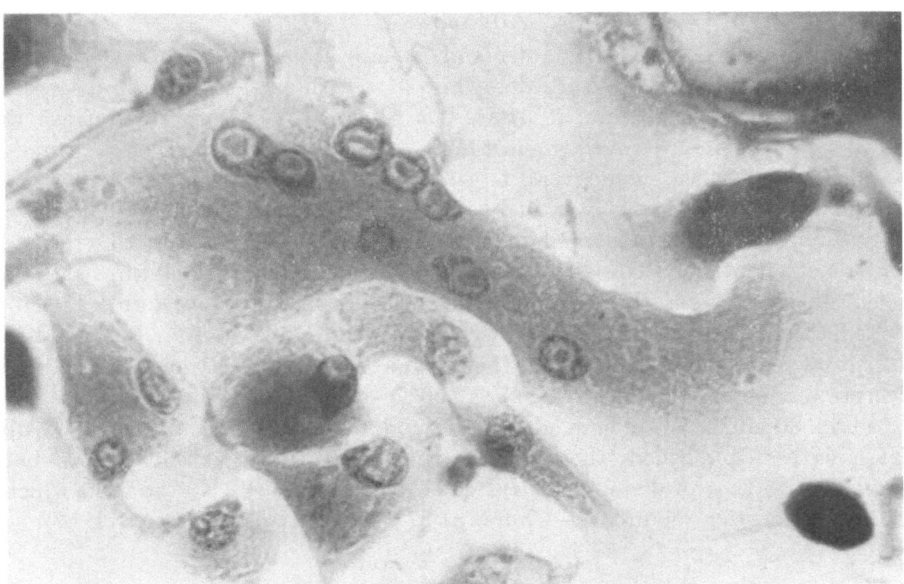

Fig. 5. Same culture as Fig. 2 but 20 days after infection. A small syncytium with numerous well-defined intranuclear inclusions. Bouin:H and E (× 680).

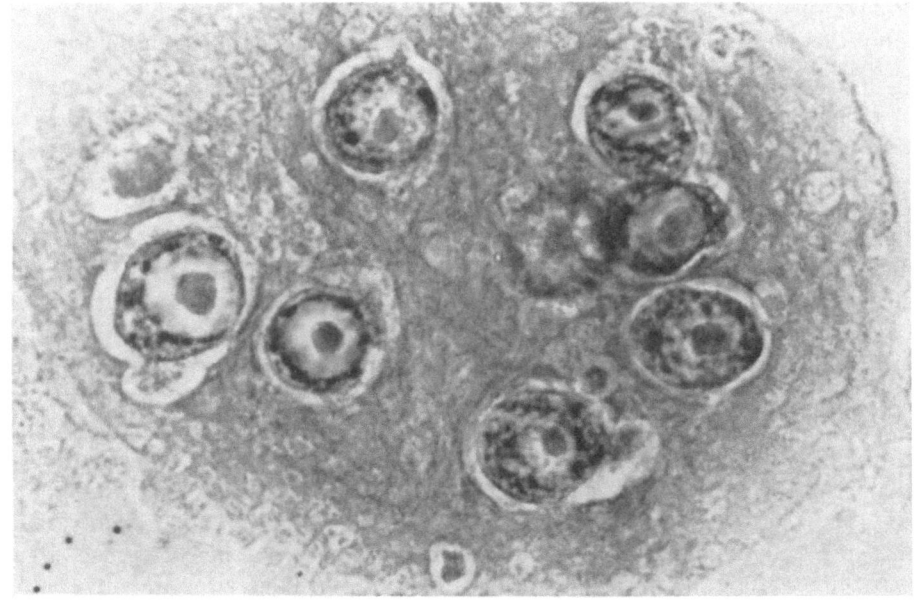

Fig. 6. Same culture as Fig. 5. Large intranuclear inclusions with partially-cleared halos and displacement of nucleoli. Bouin:H and E (× 1570).

6. Virus Multiplication in Cell Cultures

a) Adsorption

In three experiments PLOWRIGHT and FERRIS (1959a) found that 92—97% of the input virus was adsorbed within 5 to 8 hours to trypsin-dispersed BK cells in growth medium at 37°C, but the rate of adsorption during the first 2 hours was slower in two instances. Virus was adsorbed more rapidly to cells in established monolayers, 60—84% of the input virus being removed in 2 hours at 37°C. LIESS and PLOWRIGHT (1963b) reported that about 99% of the input virus was adsorbed to HeLa cell monolayers within 5 hours and further data for the RBOK and RGK/1 strains in primary BK cultures showed that 60—94% was adsorbed or inactivated within 4 hours (PLOWRIGHT, 1964b).

Using the plaque technique and BK monolayers, McKERCHER (1963) studied adsorption periods up to 3 hours but then stated "from all indications, the optimal adsorption time is about 5 hours; about 50% of plaque-forming units were adsorbed in 30 minutes". These rather inadequate observations seem to be all that have been published; they indicate that adsorption is relatively slow but comparable to that of distemper virus to chick embryo cells, a system in which maximal adsorption required 2—4 hours at 37°C (BUSSELL and KARZON, 1962).

b) Eclipse Phase

Using the RBOK strain in primary BK monolayers first progeny or eluted virus was detected at 6 hours in one experiment (PLOWRIGHT and FERRIS, 1959a). It was found in later experiments that no progeny virus had appeared by 12 or 14 hours, when employing the RBOK and RGK/1 strains at input multiplicities of about 1:35 to 1:100; progeny virus was consistently detected, however, at 20—24 hours (PLOWRIGHT, 1964b). The length of the eclipse phase with inocula of this order is probably therefore between 14 and 20 hours, but it has not been precisely determined.

c) Growth Phase

Using the RBOK strain in established BK monolayers, cell-associated virus (CAV) showed a rapid increase at the end of the eclipse phase, reaching peak levels of about $10^{4.5}$ to $10^{5.5}$ TCD_{50} per ml (fluid equivalent) after 48—72 hours (PLOWRIGHT, 1962a; 1964b). If used at a comparable input ratio (1:100) the RGK/1 strain produced maximal CAV yields of the same order in about 96 hours, smaller inocula delaying the peak to 8 days (PLOWRIGHT, 1964b). In HeLa cells infected with the RBOK strain at a multiplicity of ca. 1:1 peak CAV titres were reached after 5 days (LIESS and PLOWRIGHT, 1963b). No experiments in which the conditions approached those of a one-step growth curve have been reported.

In all the above experiments CAV fractions were prepared by detaching cells with versene (0.02%) or versene and trypsin (0.01%), pooling them with spontaneously-detached cells and disrupting them in an ultrasonic bath. It was considered at first that CAV equalled or exceeded released virus (RV) throughout the growth period in BK cells and the calculations of the number of infectious units per cell indicated rapid and continuous release of newly-formed virus (PLOWRIGHT, 1962a). In HeLa cells, however, intracellular virus was invariably 10—100 times greater than RV (LIESS and PLOWRIGHT, 1963b) and further investigations, using primary BK cells, revealed that the versene used to detach cells from the glass also removed $\gtrsim 90$—99% of the total CAV of the high-passage, attenuated RBOK strain thus causing the estimate of total CAV of be $\gtrsim 1.0-$ 2.0 $\log_{10}$ units lower than it should have been. With the low-passage RBOK and RGK/1 virulent strains a smaller proportion of the CAV was removed by versene and CAV titres exceeded those of RV except late in the growth cycle of the RBOK strain (PLOWRIGHT, 1964b).

All other published data on the growth of rinderpest virus in cell cultures refer to RV only. For instance, PLOWRIGHT and FERRIS (1959a) showed a plateau of virus production in BK cells, extending from about the 7th or 8th days to the 18th to 21st days; BREESE and DE BOER (1963), using a virulent Pendik strain at high multiplicity, found a maintained plateau of virus production from the 6th to the 15th days. At 40°C GILBERT and MONNIER (1962a; b) showed a more rapid decline in virus production after the 11th day when a bovine strain was used but not until after the 3rd week with the virus of "peste des petits ruminants". Maximal titres were generally of the order $10^{5.5}$ to $10^{6.5}$ TCD$_{50}$ per ml of fluid, but peak titres of $10^{7.5}$ TCD$_{50}$ per ml were recorded for the Pendik strain by DE BOER (1961). In all of these studies medium changes were carried out at intervals of several days and hence heat inactivation of newly-produced and heat-labile virus was not taken into account; it is probable that virus release occurs continuously over prolonged periods and that a better measure of virus production, as with measles virus, would be by harvesting fluids for titration at short intervals, for example, hourly (BLACK et al., 1959; WARREN, 1962).

7. Interferon and the Growth of Rinderpest Virus in Cell Cultures

An interferon preparation, produced in calf kidney cells infected with Sindbis virus, suppressed the growth of a small inoculum of the virulent RGK/1 strain of virus in the same cell type (PLOWRIGHT and FINTER — unpublished); CAV yields were delayed and suppressed by $0.2-1.2 \log_{10}$ units, the effect being most marked during the early stage of infection. Yields of released virus were suppressed to a greater extent ($0.6-2.0 \log_{10}$ units) throughout the period of observation (Graph I).

No explanation was obtained for the more marked suppression of the yield of released virus. It is not known what effect interferon may have on the growth of other strains of virus, particularly in producing the type of self-limiting cytopathic changes which are often seen with small inocula in established cultures, particularly of the more "resistant" cell types. The possible effects of actinomycin D on the production of interferon and rinderpest virus in cell cultures, should now be determined as for measles virus (ANDERSON and ATHERTON, 1964; MATUMOTO, 1966).

F. Pathogenesis

1. Natural Host Range

a) Domestic Animals

All domesticated cattle and water buffaloes are susceptible to natural infection with rinderpest virus and these species constitute by far the most important natural hosts (Table 6). Nevertheless, in recent years considerable attention has been paid to the possibility of widespread infections in small domestic ruminants

Table 6. *Domestic Animals Naturally Infected with Rinderpest*

Suborder	Subfamily	Genus	Specific name	Common name	References
Ruminantia	Bovinae	Bos	B. taurus	Ox, cattle	Curasson (1932, 1942)
			B. indicus	Zebu	Curasson (1932, 1942)
		Bubalus	B. indicus	Water buffalo	Curasson (1932, 1942), Anon (1966)
			B. sondaicus	Water buffalo	Curasson (1932, 1942), Anon (1966)
			B. mindorensis		Curasson (1932, 1942), Anon (1966)
		Poëphagus	P. grunniens	Yak	Curasson (1932, 1942), Cilli et al. (1951)
	Ovinae	Ovis	O. aries	Sheep	Curasson (1932, 1942)
		Capra	C. hircus	Goat	Curasson (1932, 1942)
Suiformes	Suidae	Suis	S. scrofa	Pig	Curasson (1932, 1942)

as an important factor in the maintenance of rinderpest virus in enzootic areas; the earlier literature on this subject has been reviewed by Curasson (1932; 1942), Dhanda and Manjrekar (1952) and Scott (1955b).

Since these reports, natural outbreaks of the disease have been observed in goats (Libeau and Scott, 1960; Sharma, 1965) and sheep (Johnson, 1958). In addition, serological evidence of extensive subclinical infection in goats has been obtained in several West African countries (Anon, 1966a; Zwart and Rowe, 1966); many sheep in Nigeria also showed antibodies, especially in the vicinity of a natural outbreak of the disease in cattle, when the overall proportion of serological positives in sheep rose to over 50% (Zwart and Rowe, 1966).

A disease known as "peste des petits ruminants" has been recorded since 1942 in sheep and goats in some ex-French territories of West Africa; the clinical and post-mortem findings are very similar to those of bovine rinderpest whilst immunologically, as well as in its behaviour in tissue cultures, the virus is virtually identical to that of classical bovine rinderpest (Mornet et al., 1956a, b; Gilbert and Monnier, 1962b). Cattle, apparently, are not infected by close contact with sick sheep and goats and it appears that the causal agent is a strain of rinderpest virus which has lost its capacity to infect cattle by the natural route, but spreads quite readily in small ruminants.

With regard to the natural susceptibility of yaks, the majority of information seems to be derived from episodes in zoological gardens (Curasson, 1932; 1942; Cilli et al., 1951), but Verma (1965) reported that in Nepal these animals are regularly immunised with lapinised virus vaccine, being too susceptible to allow the use of the more pathogenic caprinised virus.

Whether camels suffer naturally from rinderpest is a matter of controversy, since clinical and field reports have not been confirmed virologically. Curasson

(1932, 1942) regarded both *Camelus bactrianus* and *C. dromedarius* as susceptible and reviewed the earlier literature. SAMARTSEV et al. (1940) concluded that camels were not susceptible to the disease, but more recently DHILLON (1959) reported 17 outbreaks in India causing nearly a 50% mortality and assumed that the disease spread from camels to cattle and *vice versa*. SCOTT and MACDONALD (1962) found no serological evidence of rinderpest in camels originating from an area where the disease had recently been epizootic in many other species. Experimentally, camels can undoubtedly be infected by parenteral routes (LINGARD, 1905) and the disease can spread from cattle to camels by close contact (TAYLOR, 1967).

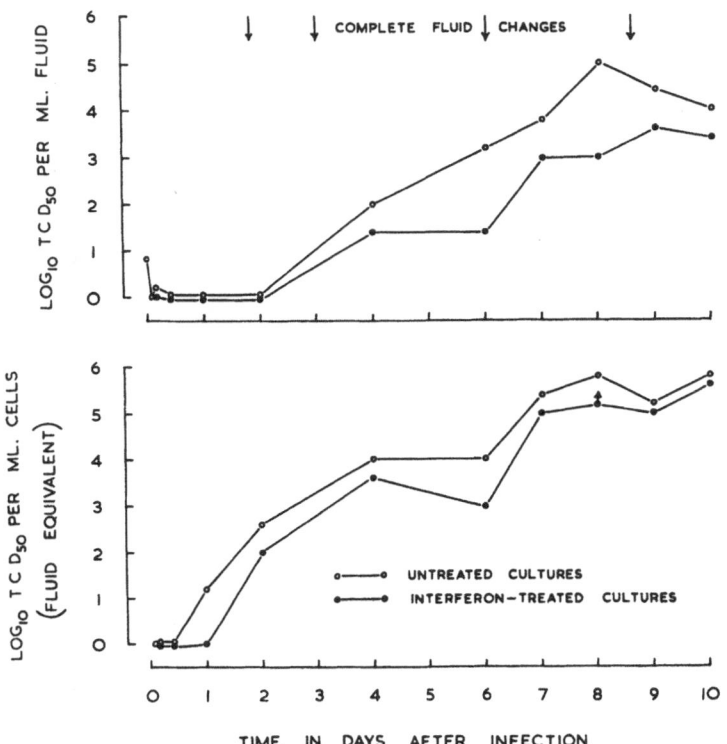

Graph I. The Effect of Interferon on the Growth of Virulent Rinderpestvirus, Strain RGK/1, in Calf Kidney Cells.

Rinderpest has long been known to affect domestic swine in Asia, especially in the south-east (CARRÉ and FRAIMBAULT, 1898; BOYNTON, 1916). Elsewhere, however, clinically-apparent rinderpest has not been reliably recorded in European-type pigs, but it has been confirmed recently that the infection can be established in them, usually in an inapparent form, by feeding infected materials or by contact with infected cattle and pigs (SCOTT et al., 1959, 1962). This is a finding of considerable epizootiological significance, since pigs could be responsible for the transfer of virus from meat offals to cattle in countries hitherto free of the disease; in fact, it has been suggested that this may have occurred in Western Australia in 1923 (ROBERTSON, 1924).

Table 7. *Freeliving or Captive Animals of the Order Artiodactyla in Which Natural Rinderpest has been Proved (After* SCOTT, *1964)*

Suborder	Family	Specific name	Common name	References
Suiformes	Hippopotamidae	Hippopotamus amphibius	Hippopotamus	PLOWRIGHT et al. (1964)
	Suidae	Phacochoerus aethiopicus	Warthog	HILSONT and BOURDEREAU (1954), WHITE (1958a)
		Sus scrofa	Wild pig	SYNTIN (1928)
Ruminantia	Giraffidae	G. reticulata	Reticulated giraffe	PLOWRIGHT (1963c)
	Bovidae	Syncerus caffer	African buffalo	THOMAS and REID (1944), GUYAUX (1951)
		Antelope cervicapra	Blackbuck	LINGARD (1905); GUPTA and VERMA (1949)
		Gazella thompsonii	Thompson's gazelle	PLOWRIGHT (1963c)[1]
		Aepyceros melampus	Impala	SCOTT et al. (1960)
		Redunca redunca	Reedbuck	CARMICHAEL (1938b)
		Ourebia ourebi	Oribi	CARMICHAEL (1938a)
		Oryx beisa	Beisa oryx	PLOWRIGHT (1961)[1]
		Damaliscus korrigum	Topi	TAYLOR (1964)[1]
		Connochaetes taurinus	Blue wildebeest	CORNELL (1934)
		Tragelaphus scriptus	Bushbuck	CARMICHAEL (1938a); CILLI et al. (1951); LIESS (1963)
		Strepsiceros strepsiceros	Greater kudu	THOMAS and REID (1944)
		Taurotragus oryx	Eland	CARMICHAEL (1933); THOMAS and REID (1944)
		Limnotragus spekii	Sitatunga	CARMICHAEL (1938b)
	Cervidae	Rusa aristotelis	Sambhar deer	GUPTA and VERMA (1949)

[1] Serological evidence only; virus not recovered.

b) Wild Animals

It is difficult to quarrel with the generalisation of SCOTT (1959b) that all and exclusively species of the Order Artiodactyla are susceptible to natural infection with rinderpest virus. Reports of the disease in members of other Orders can be definitely discounted; lists of species known more or less confidently to have been infected naturally were provided by CURASSON (1932, 1942); SCOTT (1964) and ANON (1966a). The number is constantly being extended, especially in the case of the African antelopes and it may seem a little pedantic to demand virological proof of infection for each of the many species involved in widespread epizootics of clinically-typical rinderpest occurring in former years, whether in the wild or in zoological gardens. It was seldom possible or practicable in the past to provide absolutely conclusive proof for each such species but with the advent of cheaper, more rapid methods of reliable diagnosis there is little excuse to miss any future opportunities to extend or confirm the list.

Table 7 provides details for some free-living or captive species from which the virus has been recovered in the course of spontaneous outbreaks of the disease or in which specific antigens or antibodies have been demonstrated. Table 8 lists those species in which infection has been established experimentally and Table 9, those which have been reported

Table 8. *Species of the Order Artiodactyla Experimentally Infected with Rinderpest (after SCOTT, 1964)*

Suborder	Family	Specific name	Common name	References
Suiformes	Suidae	Potomochoerus porcus	Bushpig	POULTON (1914); CARMICHAEL (1938a); WILDE (1948)
Ruminantia	Bovidae	Bibos frontalis	Gaur	ORR (1945)
		Adenota kob	Uganda kob	METTAM (1937)
		Kobus, sp.	Defassa waterbuck (?)	METTAM (1937)
		Gazella dorcas (?)	N. African gazelle	NICOLLE and ADIL BEY (1901); CURASSON (1932)
		Rhynchotragus kirkii (?)	Dik-dik	WILDE (1949)
		Cephalophus coronatus	Duiker	CURASSON (1932); DAUBNEY (1943)
		Sylvicapra grimmia	Grey duiker	CACCAVELLA (1936)
	Cervidae	Cervus sika	Japanese deer	ONO and KONDO (1923)

(?) Species not identified.

Table 9. *Species of the Order Artiodactyla which have been Reported to be Susceptible to Rinderpest (after Scott, 1964)*

Suborder	Family	Specific name	Common name	Location	References
Suiformes	Tayassuidae	Dicotyles tajacu	Peccary	Zoo	Sainte-Hilaire (1865)
	Suidae	Sus cristatus	Asiatic wild pig	Wild	Molinie (1931)
		Hylochoerus meinertzhageni	Giant forest hog	Wild	Carmichael (1938a); Guyaux (1951)
Ruminantia	Tragulidae	Tragulus sp.	Chevrotain	Zoo	Sainte-Hilaire (1865)
	Cervidae	Cervulus muntjak	Barking deer	Zoo	Sainte-Hilaire (1865); Gupta and Verma (1949)
		Blastoceros sp.	Pampas deer	Farm	Roberts (1921)
		Cervus porcinus	Hog deer	Zoo	Gupta and Verma (1949)
		Cervus axis	Spotted deer	Zoo	Gupta and Verma (1949)
		Cervus sp.	European deer	?	Bristowe (1866)
		Mazama sp.	Brocket deer	Zoo	Sainte-Hilaire (1865)
		Capreolus sp.	Roe deer	Wild	Broudin (1923)
	Bovidae	Boocerus eurycerus	Bongo	Wild	Percival (1918)
		Tetracerus quadricornus	4-horned antelope	Zoo	Gupta and Verma (1949)
		Boselaphus tragocamelus	Nilgai	Zoo	Gupta and Verma (1949)
		Strepsiceros imberbis	Lesser kudu	Wild	Theiler (1897a)
		Hippotragus equinus	Roan antelope	Wild	Percival (1918); Pecaud (1924)
		Alcelaphus sp.	Hartebeest	Wild	Pecaud (1924)
		Damaliscus sp.	Bontebok, Topi	Wild	Theiler (1897a); Pecaud (1924)

Family	Species	Common name		Reference
Bovidae	Addax nasomaculata	Addax	Wild	Curasson (1932)
	Hippotragus niger	Sable antelope	Wild	Curasson (1932)
	Connochaetes gnu	Black wildebeest	Wild	Curasson (1932)
	Raphiceros campestris	Steinbok	Wild	Theiler (1897a)
	Oreotragus oreotragus	Klipspringer	Wild	Curasson (1932)
	Bibos gaurus	Gaur or Indian bison	?	Pease (1894)
	Bibos sauveli	Kouprey	?	Anon (1966a)
	Bibos sondiacus	Banteng	?	Anon (1966a)
	Litocranius walleri	Gerenuk	Zoo	Cilli et al. (1951)
	Antidorcas marsupialis	Springbok	Zoo/Wild	Sainte-Hilaire (1865); Theiler (1897a)
	Kobus ellipsiprymnus	Waterbuck	Wild	Curasson (1932)
	Naemorhedus maritimus	Goral	Wild	Curasson (1932)
	Ammotragus lervia	Maned sheep	Wild	Curasson (1932)
	Bison bonassus	European bison	Wild	Dieckerhoff (1890)
Giraffidae	Giraffa camelopardalis	Common giraffe	Wild	Percival (1918)

to be susceptible solely because they developed clinical signs of the disease during outbreaks in the wild or in zoological gardens.

A general consideration always to be borne in mind is that strains of rinderpest virus apparently differ in their capacity to infect certain species so that, for example, only African buffaloes *(Syncerus caffer)* may be seen to be sick in some natural outbreaks, whereas other highly-susceptible species, such as eland and giraffe escape completely (KINLOCH, 1963). In enzootic areas also strains of generally reduced virulence and, probably, invasiveness occur; these produce clinically apparent infection in only one or two of many species at risk (PLOWRIGHT and McCULLOCH, 1967). Of all the species of African game animals which are probably susceptible the buffalo, eland, wildebeest, warthog, bushpig, kudu and giraffe are generally regarded as most commonly affected and have greatest epizootiological significance (CARMICHAEL, 1938a; THOMAS and REID, 1944; WILDE, 1953; PLOWRIGHT, 1963a).

2. Experimental Host Range

a) Goats

It is difficult to overestimate the importance in the past of goats as experimental hosts for rinderpest, simply because they provided the first attenuated virus for use as a vaccine and the means of its large-scale production, even in countries lacking elaborate facilities and highly-trained manpower. The size of the operations involved can be judged from figures for the numbers of cattle inoculated; thus, in India during the 10 years from 1954 about 130 million cattle received the goat-adapted (caprinised) vaccine (ANON, 1965); in E. and N.E. Africa, excluding Egypt, the total annual usage was over 8.5 million doses in 1960 (LIBEAU and SCOTT, 1960).

Historically it is interesting to note that KOCH (1897) inoculated infected cattle blood into goats and found that they exhibited only a pyrexia after an incubation period of 2 to 3 days; the virus was passaged in this species 7 times and returned to cattle after 2 and 5 goat passages; some reduction in virulence was possibly observed, as 2 of 4 cattle recovered and KOCH considered that definite, though slow, attenuation would probably be possible by repeated passage. NICOLLE and ADIL BEY (1899) found that the majority of inoculated goats succumbed with fever and progressive emaciation; some breeds were more sensitive than others. TODD and WHITE (1914) reported severe reactions in Bedouin goats given virulent cattle blood, whereas Egyptian goats showed only occasional and mild pyrexia.

SCHEIN (1917) used goats in Indochina to propagate rinderpest virus and after 172 direct goat passages noted that the virus still killed cattle; later, however, he noted signs of attenuation (SCHEIN, 1926) as did also TOPACIO (1926) after a single passage. Credit for the development and practical application of this finding to vaccine production must, however, remain with EDWARDS and his co-workers at Mukteswar in India (BROTHERSTON, 1956). The primary objective in using goats had been to produce virulent blood free of protozoa pathogenic for cattle, and therefore safe for use in the serum-simultaneous method of immunisation. Work began towards the end of 1926, the virus being first established in a pregnant goat by inoculation into the foetal side of the placental membranes; this route was adopted because EDWARDS believed that serial passage in goats or sheep was extremely difficult by conventional parenteral routes (EDWARDS, 1927a, 1930). The pregnant goat showed a temperature reaction and blood collected on the 3rd day was subinoculated to other goats by the subcutaneous route. On further goat passage the pathogenicity for this species and sheep increased (EDWARDS, 1927a). Incidentally, WALKER (1929) referred to initiating a series of 17 direct passages in goats by inoculation of the *os uteri* but gave no reason for selecting this route; he also used a single goat passage to obtain virulent virus free of bovine pathogens.

By early 1929, EDWARDS (1930) was able to report that serial direct or alternate passage for nearly 2 years in goats had produced a very striking, if fortuitous, attenuation of his virus. STIRLING (1932, 1933) employed EDWARDS' "fixed" goat virus to immunise cattle in the field and found that it could be safely used without the simultaneous administration of serum; he also observed that bovine virus from Mukteswar could be "adapted" to goats by subcutaneous inoculation in more than 50% of trials (STIRLING, 1932).

SAUNDERS and AYYAR (1936) re-investigated the effect of prolonged goat passage on the cattle virulence of an Indian strain of rinderpest virus and found evidence of attenuation after 80 passages. DAUBNEY (1948, 1949) summarised the results of passage of the Kenya RBOK strain in goats; the mortality in cattle inoculated with virus of the first 50 goat passages was 100% and the first recovery occurred at the 85th passage; from the 85th to 150th passage 15/17 cattle recovered and clinical signs were of decreasing intensity. After 250 passages, the goat-adapted virus produced approximately a 2% mortality in zebu cattle, whilst further slight attenuation may have occurred between the 250th and 400th passages, but not thereafter.

Goats of different breeds or localities or origin were reported to vary considerably in their response to inoculation with goat-adapted virus (LALL, 1947; SHARMA and RAM, 1955; HENDERSON, 1945) and some animals appeared to be completely resistant (DELPY, 1935; HENDERSON, 1945; GIRARD and CHARITAT, 1947; CILLI, 1951; SACQUET and TROQUEREAU, 1952). Whilst it is probable that some of the resistant animals had suffered previously from natural infection, an hypothesis supported by their increased frequency in older age groups, the distribution of rinderpest-neutralising antibody in Nigerian goats did not give a reliable guide to their resistance to adapted virus, since only 38% of non-reactor goats possessed antibody (ZWART and ROWE, 1966). An innate or breed resistance may, therefore, be involved as it is with some breeds of cattle (TODD and WHITE, 1914; MORNET and GILBERT, 1958).

In the first few passages rinderpest virus may spread by contact from goat to goat (EDWARDS, 1927a; BEATON, 1930; ZWART and MACADAM, 1967a) but after more prolonged adaptation it does not do so (SEETHARAMAN, 1948; DAUBNEY, 1949; CILLI, 1951; SACQUET and TROQUEREAU, 1952). Experimentally, infection with adapted virus is usually established by the subcutaneous route but DAUBNEY (1949) reported that 96 serial passages had been effected by the intratracheal route, in an unsuccessful effort to increase the pneumotropic properties of the virus; scarification of the skin has also been used (SHARMA and RAM, 1955).

Goats inoculated with adapted (caprinised) virus usually show a temperature reaction on the 2nd day following and some may develop cough, dyspnoea or diarrhoea. The mortality is usually very high, 90—95% according to SAUNDERS and AYYAR (1936) and DAUBNEY (1949) and is associated with gastro-enteric or pneumonic lesions. The latter, however, whilst frequent in early passages (EDWARDS, 1927a; DAUBNEY, 1949) diminished later and may, in fact, have been due to intercurrent infections.

b) Rabbits

The importance of rabbits as laboratory hosts for rinderpest virus rested mainly, as with goats, on their use in the production of attenuated live vaccines but, in addition, they attained for a time considerable prominence in the performance of virus-neutralisation tests and are still employed for both these purposes and in the production of hyperimmune sera for use in complement-fixation and agar-gel diffusion tests. There are already several reviews of the

literature on infection of rabbits, including those of CURASSON (1942); CHENG
and FISCHMAN (1949); BROTHERSTON (1957); MORNET and GILBERT (1958) and
SCOTT (1964). The following outline owes much to these authors.

Unsuccessful attempts to infect rabbits with bovine virus were recorded
by NICOLLE and ADIL BEY (1899); McKINLEY (1928); MORCOS (1931) and
PHILIPPE (1939). EDWARDS (1927a) investigated rabbits primarily as potential
virus donors for the serum-simultaneous method of immunisation; he started
in 1922 by inoculating large doses of virulent ox blood intravenously and observed
a mild, transitory pyrexial response which on passage became more pronounced
and lasted from the 2nd or 3rd to the 5th or 6th days. The virus was transferred
at intervals of 2—7 days for a period of 14 months but never became safe to use
by itself as a vaccine for cattle.

HORNBY (1926a); JACOTOT (1930, 1932a) and DAUBNEY (1937a) could only
demonstrate persistence of the virulent virus in rabbits, at most, during 3 passages. INOUE (1934) passaged 3 strains of virus 30 to more than 50 times and
with one stock laboratory strain was unable to show attenuation for cattle;
CEBE and PERRIN (1935), on the other hand, were more fortunate in noting
attenuation after 36 and 46 rabbit passages of two Indochinese strains. BAKER
(1946) and CARTER and MITCHELL (1958) resorted to alternate cattle-rabbit
passages to adapt bovine and chick embryo-passaged strains to rabbits. HADDOW
and IDNANI (1947) were successful in establishing one field strain of high virulence
in rabbits, but failed with two other bovine strains; after 55 rabbit passages
they noted some loss of cattle pathogenicity. IYER and SRINAVASAN (1954)
reported on the adaptation of another bovine strain to rabbits and found it
sufficiently attenuated for use as a vaccine in buffaloes.

The most important adaptation of rinderpest virus to rabbits was carried out
by NAKAMURA and his colleagues in Korea. In 1938 they reported that rabbits
were infected by the intravenous or subcutaneous inoculation of virulent bovine
blood, preferably by the former route; the virus could be passaged in series and,
whilst during the first five transfers only 83% developed varying degrees of pyrexia
and 18% showed specific histological lesions; in the next 5 passages 93% developed fever and 86% definite lesions (NAKAMURA et al., 1938; FUKUSHO and NAKA-
MURA, 1940). This was a distinct advance since EDWARDS (1927a); INOUE (1934),
JACOTOT (1930) and CEBE and PERRIN (1934) had all failed to observe any definite
lesions and the last two groups of workers did not even detect pyrexia.

FUKUSHO and NAKAMURA (1940) found after 3—4 serial passages that the lymphoid follicles of PEYER's patches became swollen and necrotic and from the 3rd
day onwards were visible through the serosa as aggregations of whitish nodules;
the lesions were best developed by the 5th—6th days after inoculation and thereafter receded. Similar changes were seen in the appendix, tonsilis caecalis major,
sacculus rotundus and mesenteric lymph nodes and, in addition, the spleen was
enlarged; these lesions, together with the consistent fever, provided a ready
means for identification of the rabbit-adapted or "lapinised" virus of the NAKA-
MURA III strain. The only likely confusion was with infection due to *Salmonella
typhimurium,* easily differentiated by bacteriological examination (CHENG and
FISCHMAN, 1949).

The pyrexia due to infection with the NAKAMURA III strain is typically sudden in onset, appearing after 36—48 hours with large doses of virus but delayed by smaller ones; peak temperatures of $\leq 105°$ F. were attained within hours of onset and persisted for 2—3 days. Rabbits were usually infected intravenously but the subcutaneous, intramuscular (CHENG and FISCHMAN, 1949), intraperitoneal (BROTHERSTON, 1951 a) or intracerebral routes (SCOTT and RAMPTON, 1962) gave essentially similar results. Rectal or oral instillation of virus caused infection (MUMATSEY and KOYAMU, 1945) but the dose required was considerably greater (SCOTT and RAMPTON, 1962). Contact infection did not occur (CHENG and FISCHMAN, 1949; BROTHERSTON, 1951 a; SCOTT and RAMPTON, 1961). Rabbits of all breeds and ages were susceptible but there were superficial indications that randomly-bred animals 4—6 months old were most suitable for virus propagation and that pregnant animals should be avoided (see BROTHERSTON, 1951 a).

The mortality in rabbits infected with bovine or low-passage strains of virus was low and there were no distinctive lesions (EDWARDS, 1927 a; JACOTOT, 1930). CHENG and FISCHMAN (1949) also reported that mortality was low in rabbits inoculated with the NAKAMURA III strain after more than 630 serial passages but if complications, such as coccidiosis, were present the mortality might rise to 60%. NAKAMURA obtained a mortality of 87% with virus of 751 to 900 serial passages (MORNET and GILBERT, 1958) death taking place on the 3rd to 9th days. SCOTT (1959 c), however, reported a mortality rate of 96 ± 2% in Kenya, the death time being inversely related to the virus dose. Clinical signs noted in some infected rabbits included rapid respiration (BAKER, 1946); anorexia, general malaise and occasionally diarrhoea (BAKER, 1946; CHENG and FISCHMAN, 1949; BROTHERSTON, 1951 b); nevertheless, MORNET et al. (1953) considered these signs to be rare.

The proliferation of virus in the rabbit has not been studied in detail. According to NAKAMURA et al. (1938) and MORNET et al. (1955) it appeared in the blood 10—15 hours after intravenous inoculation and the highest concentrations of virus (c. $10^7—10^8$ MID per gramme) were found in the mesenteric lymph nodes (NAKAMURA, 1941; 1957 b) or appendix (SCOTT, 1954; MORNET and GILBERT, 1958). High titres were also recorded in other intestinal aggregations of lymphoid tissue, such as the sacculus rotundus (SCOTT, 1954). Peak titres in the appendix were of the order of $10^{7.7}$ MID per gramme (SCOTT, 1954; MORNET and GILBERT, 1958), and spleen titres were 10^6 to 10^7 MID per gramme (NAKAMURA, 1957 b; SCOTT, 1954); the blood titre was only about 10^4 or 10^5 MID per ml (SCOTT, 1954; MORNET and GILBERT, 1958). Most of the virus in the blood was associated with the leukocytes (NAKAMURA, 1941) and persisted for 11 days (MACOWAN, 1956), 15 days (MORNET et al., 1955) or even in one case 22 days (CHENG and FISCHMAN, 1949). Virus in the urine was detected by CHENG and FISCHMAN (1949) but not by SCOTT (1954).

So far as its cattle pathogenicity was concerned, the Nakamura III strain of virus was still lethal for highly-susceptible Korean calves after 150 rabbit passages and it still caused occasional deaths after 325 passages (NAKAMURA and KURODA, 1942). The 350th passage caused only slight reactions in the less sensitive Mongolian cattle (ISOGAI, 1944; MUMATSEY and KOYAMU, 1945) but after more than 800 passages still caused about a 30% mortality in Japanese Black cattle (FUKUSHO and FURUYA, 1953; NAKAMURA and MIYAMOTO, 1953).

c) Dogs

Morcos (1931) produced mild pyrexia in dogs by feeding rinderpest-infected meat or by injecting virus but no conclusive proof of infection was obtained, although inoculated dogs died after a prolonged illness. A most interesting observation was that of Polding and Simpson (1957) who reported that 4 puppies given multiple injections of rinderpest virus developed neutralising antibodies. Scott and Brown (1958) found similar antibodies in 4 dogs which had been inoculated once with the RBOK strain of virus and from which the virus had been recovered in cattle. Polding et al. (1959) confirmed that the same strain of virus could be recovered from dogs 4 days after inoculation and that these animals developed high-titre neutralising antibody; no clinical reaction was, however, observed. Dogs inoculated with lapinised rinderpest virus had developed high-titre neutralising antibody 3 weeks later (Mornet et al., 1960).

DeLay et al. (1965) inoculated 13 dogs with c. 10^6 cattle ID_{50} of RBOK strain rinderpest virus and detected no clinical reactions; 12/13 developed complement-fixing antibodies for rinderpest and 9 also had low-titre neutralising antibody at 40 days after inoculation.

d) Ferrets

Ferrets have been inoculated with lapinised (Goret et al., 1957, 1960b) and virulent bovine strains of rinderpest virus without the production of any clinical abnormalities (Mornet et al., 1960). Many of them were subsequently found to be immune to challenge with the immunologically-related virus of canine distemper but rinderpest virus was not recovered from them and rinderpest-neutralising antibodies were not demonstrated in their sera. The immunity to distemper challenge was complete all 11 days after inoculation and lasted at least 11 months (Goret et al., 1960b).

e) Monkeys

Curasson (1942) reported that he had failed to infect 3 species of monkeys of the genus *Cercopithecus* and one species of baboon; he did not state whether he had attempted to recover virus. DeLay et al. (1965) inoculated 9 monkeys *(Cynomolgus)* with at least $10^{5.3}$ cattle ID_{50} of virulent rinderpest virus (RBOK strain). Virus recovery was not attempted but 8/9 had developed complement-fixing antibody and 5/9 rinderpest-neutralising antibody when examined 44 days later; neutralising titres were minimal and remained so after challenge with measles virus. It is still doubtful, therefore, whether monkeys are susceptible to infection with rinderpest virus.

f) Other Rodents

Unsuccessful attempts to infect guinea pigs were recorded by Nicolle and Adil Bey (1899); Woolley (1906); Morcos (1931) and Scott and Witcomb (1958a). Malfroy (1927) reported equivocal results, whilst Inoue (1934) and Baker et al. (1946) were able to maintain the virus through 7 or 2 passages respectively in guinea pigs; the latter authors demonstrated neutralising antibodies but only observed occasional slight temperature reactions as a result of infection. Guinea pigs given rinderpest culture vaccine intramuscularly or intraperitoneally regularly develop neutralising antibody but it is not known whether the virus proliferates in them (PLOWRIGHT, 1963 — unpublished).

The giant rat, *Cricetomys gambianus*, frequently developed fever following inoculation of virus, the blood being virulent for cattle (CURASSON, 1942) and the Mongolian suslik *(Citellus mongoliscus ramosus)* was successfully employed for 24 serial passages of the virus by INOUE et al. (1930). In the latter species pyrexia and deaths were reported only in early passages.

Syrian hamsters have been infected successfully with a lapinised-avianised (NAKAMURA et al., 1957c) and bovine (RBOK) strain of virus (SCOTT and WIT-COMB, 1958a). The latter authors passaged the virus 140 times in hamsters without observing any clinical abnormalities, no matter which parenteral route of inoculation was employed, while infection by the intranasal, conjunctival and oral routes failed. The virus was considerably attenuated for cattle, in which the mortality rate dropped from 76% to 6%. Neutralising antibodies developed in the infected hamsters (SCOTT and BROWN, 1958; PLOWRIGHT and FERRIS, 1959a).

Early attempts to adapt rinderpest virus to serial passage in the brains of mice failed (e.g. DAUBNEY, 1937b; CARMICHAEL, 1938b; CURASSON, 1942). SCOTT and WITCOMB (1958a) succeeded, however, in propagating the hamster-adapted and a bovine strain of virus in mice, which were inoculated by the subcutaneous, intraperitoneal or intracerebral routes. The virus was recoverable in cattle from spleen harvested on the 4th to 14th days post-inoculation; the virus of hamster origin caused a reduced mortality (20%) in cattle, whereas the bovine strain retained the capacity to kill 78%. Although infected mice showed a high mortality this was considered to be due to the activation of another latent infection.

IMAGAWA (1965) inoculated the culture-adapted, lapinised virus of ISOGAI (1961) into day-old mice of the CFW strain, using the intracerebral route; mice became sick at first after 14 days but after 9 passages the incubation period was reduced to 4—6 days whilst the mortality rose to 100% by the 6th passage. The titre of 10% mouse brain suspensions was $10^{4.8}$ to $10^{5.8}$ ID_{50} per ml and the virus could be easily recovered in embryonic kidney cells. The clinical signs observed in suckling mice included running and jumping, apathy and paralysis. Adult mice could be infected occasionally but the virus could not be passed in them.

Rats could not be infected with rinderpest virus (MORCOS, 1931; CURASSON, 1932, 1942) but CURASSON (1942) reported success with a hyrax *(Procavia* sp.). CARMICHAEL (1939) was unable to infect hedgehogs and WILDE (1948) failed to recover rinderpest virus from young procupines inoculated subcutaneously.

g) Marsupials

Two opossums and two kangaroos which were inoculated or drenched with virulent blood did not develop signs of disease attributable to the virus (ROBERTSON, 1924).

h) Embryonated Eggs

Much of the early work on chick embryos as a host for rinderpest virus was reviewed by BROTHERSTON (1958).

KUNERT (1938), working in Tanganyika with a virulent laboratory strain and a field strain of virus, inoculated blood or spleen suspensions onto the chorio-allantoic membrane (CAM) of 8-day embryonated chicken eggs. In some cases 5—7 days later he observed oedema of the CAM and occasionally haemorrhagic or necrotic foci; these changes were transmissible and, in one series, persistence of virus was proved for at least 5 days — by the inoculation of calves. SHOPE et al. (1946a) working on Grosse Isle from 1942 onwards were able to report by 1943 that the RBOK strain of virus could be serially propagated in the CAM of 10-day chick embryos. Incubation was normally at 37.5—38° C and membranes were harvested at 3 days; virus was absent from extra-embryonic fluids and embryos, unless 8-day eggs were incubated for 5 days post-inoculation.

It was later reported (SHOPE et al., 1946b) that virus passaged 8—12 times on the CAM caused regular infection of the embryo if inoculated into the yolk sac of 6—7 day eggs; this extension of the virus at first required 48—72 hours but later only 24 hours. Many infected embryos died on the 2nd to 4th days but this was not a constant phenomenon. Some attenuation for cattle took place on passage on the CAM but the process was faster and more regular when inoculation was into the yolk sac (JENKINS and SHOPE, 1946) and virus of the 41st yolk sac passage produced only a mild illness in calves. HALE and WALKER (1946) mentioned that goat-adapted and Indian strains of virus would not adapt to the yolk sac route after 19 and 30 passages respectively on the CAM; the former did so after 37 membrane passages, however, whilst another bovine strain of N. African origin would not grow on the CAM at all.

CHENG, CHOW and FISCHMAN (1949) also reported their failure to grow an Egyptian and two Chinese virulent strains on the membranes of 10-day eggs, but their success on two occasions with the RBOK strain. In one of the latter transmission series the virus was adapted to the yolk sac after 17 membrane transfers but in the other yolk sac adaptation failed after 48 such passages. HUDSON (1947) also succeeded with the same strain in two series of CAM passages but growth in the yolk sac was not demonstrable (HUDSON and DANKS, 1949); one series, "K", was attenuated for cattle from the 39th passage onwards, the other, "L", after 45 passages, i.e. yolk sac passage was not essential for attenuation. DAUBNEY (1951); CARTER (1952) and McKERCHER (1957) also found that the RBOK strain of virus could be adapted to eggs and the latter author determined that attenuation in 5 series occurred between the 30th and 43rd CAM passages; again, therefore, yolk sac passage was not necessary for attenuation but did increase the potential yield of vaccine from eggs.

NAKAMURA et al. (1947) inoculated 12—14 day embryos intravenously with lapinised virus of the Nakamura III strain, which they wished to attenuate further and found that it multiplied in the embryonic spleen but not in sufficient quantity to allow further egg transfers. They then used the same method for a virulent bovine strain which they were able to transfer 3—14 times in 4 passage series; apparent attenuation for cattle was noted once after 10 passages. WALKER (1947) inoculated calf spleen infected with the lapinised strain of BAKER (1946) into the yolk sac of eggs and was able to passage the virus at least 29 times by this method. NAKAMURA and MIYAMOTO (1953) finally succeeded in adapting lapinised virus to growth in eggs by alternating passage on the CAM or by the intravenous route on the one hand and in rabbits on the other; they were eventually able to maintain the virus indefinitely by direct intravenous passage in eggs. During the adaptation period the virus decreased in pathogenicity for rabbits (the mortality dropping to 20%) and also for cattle; hypersusceptible Korean and Japanese Black calves now showed only a slight to moderate thermal reaction, developed antibodies and resisted challenge inoculation (NAKAMURA and MIYA-MOTO, 1953; NAKAMURA, 1957a).

FURUTANI et al. (1957a) also adapted the Nakamura III strain of lapinised virus to passage by the intravenous route in chick embryos; they followed a similar procedure to NAKAMURA and MIYAMOTO (1953) noting, in addition, embryo deaths and reddening and swelling of embryonic spleens. Their strain was called AKO. In Kenya, WITCOMB et al. (1962) found that LA virus given by the yolk sac route also produced a high mortality of embryos, viz.: $51.1 \pm 14.1\%$. NAKAMURA et al. (1954) found maximal titres (10^5 to 10^6) in the spleen, FURUTANI et al. (1957b) in the blood, yolk sac and CAM.

FUKUSHO successfully carried the virulent Fusan bovine strain of virus through 72 membrane passages and failed in 10 attempts to adapt it to the yolk sac route; he eventually succeeded in this objective by using cattle spleen in-fected with the membrane-passaged virus as inoculum. This yolk sac strain was designated BA but was still too virulent for use in Japanese calves 158 passages later (ISHII and TSUKUDA, 1952). It produced a very high embryo mortality ($89.9 \pm 4.1\%$) when inoculated into the yolk sac of 5-day embryos; deaths occurred mainly between the 6th and 12th days with a peak on the 9th day (WITCOMB et al., 1962).

i) Chickens

KYLASAMAIER (1931) reported that domestic fowls died as a result of inoculation of virulent bovine liver or blood but the virus could not be demonstrated in fowl tissues by subinoculation to cattle. RAJAGOPALAN (1937) also failed to recover rinderpest virus from chickens inoculated with virulent ox blood. SHOPE and GRIFFITHS (1946) found that if chick embryos were inoculated into the yolk sac a small proportion hatched normally and virus persisted in their brains and solid viscera for 5, not 8, days; they also developed antibodies. Day-old chicks could also be infected intracerebrally but not by other routes and the virus could not be passed serially. BAKER and GREIG (1946) found that day-old chicks could be infected by the intravenous inoculation of embryo-adapted virus and that they also developed antibodies. It was suggested that the production of antibodies in chicks could be useful in evaluating chick-embryo vaccines.

3. Development of the Virus in Cattle

a) The Site of Entry

Cattle can be infected experimentally by any parenteral route of inoculation — subcutaneous, intravenous, intraperitoneal, intratracheal (NICOLLE and ADIL BEY, 1899; HORNBY, 1926b); intradermal (HORNBY, 1926b; CURASSON, 1942; SCOTT,

1952) and intracerebral (NICOLLE and ADIL BEY, 1901). Rectal deposition of virus was not successful according to HORNBY (1926b) and mixed results were obtained by vaginal deposition (TODD and WHITE, 1914). Infection took place readily with nasal discharges administered by conjunctival instillation (TODD and WHITE, 1914) but HORNBY (1926b) succeeded in only one of two experiments.

The virus is generally regarded as incapable of passing through the intact skin or other stratified squamous epithelia (HORNBY, 1926b); this presumably accounts for frequent failures to infect by feeding or drenching with highly virulent material (TODD and WHITE, 1914; SCHEIN and JACOTOT, 1925; HORNBY, 1926b; MAURER, 1963); by giving naturally-contaminated water to drink (TODD and WHITE, 1914) or by intrarumenal inoculation (HORNBY, 1926b). Pigs, as opposed to cattle, appear to be easily infected by eating contaminated food or by drenching them with infective materials (BOYNTON, 1916; SCOTT et al., 1962). If there are discontinuities of the skin epithelium, as caused by scarification, the skin may allow passage of the virus (HORNBY, 1926b; VAN SACEGHEM, 1923; CURASSON, 1942), albeit in an irregular manner (TODD and WHITE, 1914).

Infection takes place readily via the upper respiratory tract, as by nasal swabbing with the excretions of infected animals or by intranasal instillation of virus (TODD and WHITE, 1914; HORNBY, 1926b; HALL, 1933). Nasal swabbing was, in fact, often used as a method of infection which eliminated the danger of concurrent transmission of protozoal pathogens. Infection took place when cattle were exposed at a distance of no greater than 6 feet from the heads of other cattle infected with one highly virulent strain; transmission was presumably mediated by infective aerosols (IDNANI, 1944). TODD and WHITE (1914) failed, however, in similar experiments in which animals were kept at a distance of 1 metre only, with wire netting to prevent closer contact. PROVOST (1958) had no difficulty in infecting cattle with experimental aerosols of a virulent strain of virus.

LIESS and PLOWRIGHT (1964) and PLOWRIGHT (1964a) used intranasal instillation to infect cattle with a virulent field strain of virus derived from a giraffe (RGK/1). The animals consistently showed temperature reactions on the 3rd to 5th days following inoculation with $10^{3.0}$ to $10^{5.1}$ TCD_{50}; two or three animals were killed every day from the first to the tenth and also on the 12th, 14th, 15th and 16th days following infection and 20—21 tissues were harvested separately for titration of their virus content (PLOWRIGHT, 1964a). No primary multiplication of virus was detected in the mucosa of the nasal turbinates but virus was demonstrable after 24—48 hours in the pharyngeal and/or mandibular lymph nodes as well as in the palatal tonsil; viraemia was detected from the 2nd to 3rd day onwards, i.e. one to two days before the appearance of pyrexia.

TAYLOR et al. (1965) investigated the route of infection following 24 hours contact with excretor cattle and found that virus first localised in all of 7 cases in the pharyngeal lymph node, in 3 cases also in the mandibular node and in 4 animals in the palatal tonsil; no virus was found in the mucosae of the turbinates or dorsal pharynx. In 3 of the 7 cattle virus was also present, and to the highest titre, in the bronchial and costocervical lymph nodes but no infectivity was detected in the lung and bronchial or tracheal mucosae.

These data were interpreted to mean that natural infection probably took place invariably via the upper respiratory tract, and also sometimes via the lower respiratory tract; primary foci of proliferation in the mucosae were not found and it was suggested that the infecting virus was carried through respiratory mucosae without proliferating there.

b) The Incubation Period

Experimentally this lasts for 2—9 days, depending on the strain, route of administration and dosage administered; in cases acquired by contact the incubation period is sometimes stated to be 3—9 days (SCOTT and BROWN, 1961; MAURER, 1963), but this again must be strain dependent as recent E. African isolates required 10—15 days (ROBSON et al., 1959; PLOWRIGHT, 1963a), or 8—11 days (LIESS and PLOWRIGHT, 1964; PLOWRIGHT, 1964a) to produce disease in animals exposed by contact. The data of TAYLOR et al. (1965) indicate that incubation periods following severe contact exposure to the RGK/1 strain ranged at least from 4—5 up to 11 days.

Virus usually appears in the blood 1—2 days before the onset of fever (SCOTT, 1955a; MACOWAN, 1956; PLOWRIGHT, 1963a; LIESS and PLOWRIGHT, 1964 and PLOWRIGHT, 1964a) and has generalised throughout the body by the end of the incubation period; in particular, virus proliferation is established in both superficial and visceral lymph nodes, in the spleen and bone marrow, in the mucosae of the upper respiratory tract and in the lung and mucosae of the gastrointestinal tract from abomasum to colon (PLOWRIGHT, 1964a; TAYLOR et al., 1965). No virus proliferation has probably begun at this stage in either the nasal mucosa (PLOWRIGHT, 1964a) or in the susceptible epithelium of the tongue in areas free of lymphoid nodules (TAYLOR et al., 1965).

c) The Prodromal Phase

This term refers to the time between the onset of pyrexia and the first appearance of mucosal lesions in the mouth (PLOWRIGHT, 1964a, 1965). It lasts about 3 days on average, varying from 2—5 days, and is characterised by the attainment and maintenance of very high titres of virus in the lymphopoietic tissues, especially in the cephalic lymph nodes, tonsils and haemal nodes but also in the mucosae of the alimentary tract. Virus also begins to proliferate during this phase in the nasal mucosa, reaches considerable levels in the lungs and less in the liver. The kidney, myocardium and brain do not regularly support virus growth. The titres of virus during this "plateau" phase reached the following mean levels in different tissues:

(i) Strain RBOK (Virulent)

Lymphoid tissue, abomasal mucosa $10^{6.0}$ ID_{50}/g; blood $10^{4.5}$ ID_{50}/ml; bone marrow $10^{1.0}$—$10^{2.0}$ ID_{50}/g (SCOTT, 1955a).

(ii) Strain RGK/1

Cephalic and superficial lymph nodes, palatal tonsil, haemal nodes ca. $10^{7.0}$—$10^{8.0}$ TCD_{50}/g; spleen, visceral lymph nodes, lung, tongue mucosa ca. $10^{6.0}$—$10^{6.5}$ TCD_{50}/g; gastro-intestinal mucosae ca. $10^{4.5}$—$10^{6.0}$ TCD_{50}/g; nasal mucosa, bone marrow, liver ca. $10^{4.0}$—$10^{4.5}$ TCD_{50}/g; blood $10^{2.3}$—$10^{3.0}$ TCD_{50}/ml (LIESS and PLOWRIGHT, 1964; PLOWRIGHT, 1964a).

These figures should be compared with earlier estimates for virulent strains obtained by methods which were almost certainly less accurate and founded on very few observations, usually incidental to the production of inactivated tissue vaccines. Thus, JACOTOT (1931a) found that the highest titre of virus (1/300,000) occurred in the abomasal mucosa, with lymph nodes and tonsil seldom exceeding 1/50,000. WALKER et al. (1946b) found that the maximum quantities of virus of the RBOK strain were found in the lymph nodes, spleen and lungs (1/10,000). All these workers titrated material in cattle or goats. Somewhat lower levels were recorded by FURUYA and FUKUSHO. They employed rabbits for titrations and these were later challenged with the NAKAMURA III strain of lapinised virus; it was contended that the virulent virus produced a silent but immunising infection (NAKAMURA, 1957a).

In cattle infected with mild field strains recently isolated in E. Africa, PLOW-RIGHT (1963a) found that the peak mean blood titre, estimated in tissue cultures, reached only $10^{1.94}$ on the 4th day of the disease. SCOTT (1959a) found that goat-adapted and lapinised vaccine strains reached peak mean blood titres of $10^{3.5}$ and $10^{1.0}$ on the 2nd and 3rd days of fever respectively. TAYLOR and PLOWRIGHT (1965) could only re-isolate culture-attenuated virus from leukocyte fractions of cattle blood on the 4th to 8th days following inoculation and it was not present in sufficient quantities to titrate. Nevertheless, the mean titres in lymph nodes, tonsils, PEYER's patches and haemal nodes reached c. $10^{4.0}$ to $10^{5.0}$ TCD_{50}/g by the 7th day. It may be dangerous to attribute too much significance to the quantity of virus circulating in the blood, since this is obviously dependent on "overspill" from lymphoid and other tissues in which the virus proliferates and the level of viraemia may not always be an accurate reflection of the degree of local virus proliferation. Nevertheless, it does appear that the quantity of virus which spills over into the blood is directly proportional to the pathogenicity of the strain for cattle. The distribution of virus multiplication in various tissues gave a more striking index of pathogenicity; thus the virulent RGK/1 strain multiplied to high titre in the mucosae of the respiratory and alimentary tracts, whereas the culture-attenuated strain could not be recovered from these situations at any time and was completely apathogenic (PLOWRIGHT, 1964a; TAYLOR and PLOWRIGHT, 1965).

d) The Mucosal and Convalescent Phases

These phases, characterised clinically by the appearance of mouth lesions and diarrhoea, followed by their recession in animals which recover, begin with high virus titres in all major sites of proliferation and in the blood. The amount of virus begins to decline rapidly, however, with the first appearance of antibody, on the 5th or 6th days of the disease in the case of the RGK/1 strain; the cephalic lymph nodes which show the first viral proliferation are also those in which the decline first begins, while the alimentary mucosae, respiratory tract tissues and lympho-epithelial structures such as the tonsils are those which retain virus longest (PLOWRIGHT, 1964a).

SCOTT (1955a) found that cattle infected with the virulent RBOK strain showed a viraemia for 10 days, whereas virus persisted in the spleen and lymph nodes over a period of 14 days after the onset of fever; caprinised and lapinised strains of virus disappeared from the blood after 9 and 8 days respectively, but persisted for longer in the spleen. PLOWRIGHT (1963a) found that mild strains of virus could be recovered from the blood of some cattle up to the 2nd or 3rd day after the subsidence of fever. The mean duration of viraemia being 6.35 days (range 3—11 days). The RGK/1 virulent strain was recovered from only one of 84 tissues from 4 animals killed on the 10th to 12th days after the onset of fever, and the single isolation was from emphysematous lung; with this strain viraemia lasted 2—8 days (mean 6.6 days) (PLOWRIGHT, 1964a).

The persistence of virus in the blood and other tissues of cattle with appreciable levels of circulating antibody (PLOWRIGHT, 1964a) can be explained by its association with the leukocytes. It has been known for about 70 years that serum or plasma from virulent ox blood is not infective (SEMMER, 1896; THEILER, 1897b; KOLLE, 1899). It was also shown very early that the infectivity in blood was very largely associated with the leukocytes and not with the erythrocytes (TODD and WHITE, 1914; SCHEIN, 1917; HORNBY, 1928; DAUBNEY, 1928). More recently it was found that leukocyte fractions prepared from infected blood by several cycles of differential centrifugation and devoid of polymorphonuclear leukocytes, had a high infectivity; fractions consisting almost entirely of platelets and with no leukocytes contained at most a trace of virus (PLOWRIGHT — unpublished observations). Hence, it can be concluded that the infectivity is certainly associated with the mononuclear cells, perhaps to the exclusion of the other leukocytes and the platelets.

It is generally accepted that animals recovered from the disease do not remain virus carriers and that virus inoculated into immune animals disappears immediately and permanently (SCOTT, 1955a). However, PLOWRIGHT and TAYLOR (1967) found that two animals vaccinated successfully with culture-attenuated virus 27 and 37 months previously, did develop a viraemia following challenge; although there was no clinical reaction whatsoever; the antibody level in these two cattle had fallen very low. JACOTOT (1929) noted the absence of clinical reaction following challenge inoculation of cattle protected by serum given simultaneously or by previous administration of inactivated vaccine, but such cattle were not regarded as retaining the virus for prolonged periods (JACOTOT, 1931b). More recently PLOWRIGHT (1963a) recorded viraemia of 4—6 days duration in two serologically-negative cattle which were inoculated with a mild strain of virus and subsequently acquired neutralising antibody and resisted challenge inoculation; hence clinically-silent infections can occur in cattle not previously exposed to virus.

Both CURASSON (1921) and JACOTOT (1931c) reported that they had detected virus in the vaginal secretions of cows which had aborted 5—12 weeks previously; JACOTOT (1931c) also recovered virus from 2 foetuses aborted 21 and 33 days after sero-vaccination, and CURASSON (1926) from abomasal lesions of a calf 34 days after apparent recovery. DATTA and RAJAGOPALAN (1932) recovered virus from the spleen of a "chronic" case of rinderpest 74 days after inoculation. DELPY (1928) affirmed that the blood of "chronic" cases could remain infective for as long as 38 days and GIBBS (1933) claimed virus recovery from 4/7 animals which had recovered up to 177 days previously; these reports, and others like them, should be regarded with great circumspection.

e) Excretion of Rinderpest Virus by Infected Cattle
(i) Nasal Secretions

According to TODD and WHITE (1914) the nasal secretions were invariably virulent on the 6th to 9th days after infection but not on the 4th day. HORNBY (1926b) found that the nasal discharges were infective by the 2nd day of pyrexia or within 5 days of infection but ceased to become so 2—3 days after the remission of fever. LIESS and PLOWRIGHT (1964) found that nasal excretion of virus first became detectable 1—2 days before fever but the nasal excretion rate rose gradually to a maximum of 87.5% on the 4th day of fever and thereafter declined so that it was not found after the 9th day; the quantity of virus reached a maximum of $10^4 - 10^5$ ID_{50} per swab between the 3rd and 7th days of pyrexia.

(ii) Urine

HORNBY (1926b) and HALL (1933) found that virus did not appear in the urine until the 2nd or 3rd day of pyrexia, whereas CURASSON (1932) first detected it on the day preceding fever, and TODD and WHITE (1914) and LIESS and PLOWRIGHT (1964) on the first day. Viruria ceased 2 days after the temperature returned to normal (HORNBY, 1926b; HALL, 1933) or in the 9th day following onset of pyrexia (CURASSON, 1932; LIESS and PLOWRIGHT, 1964). Contrary to some opinions (e.g. EDWARDS, 1930) urinary excretion is not of predominant importance; the excretor rate rose to a maximum of 62.5% on the 7th day of illness, the highest mean titre ($10^{1.7}$ TCD_{50}/ml) being attained on the 6th day (LIESS and PLOWRIGHT, 1964).

(iii) Faeces

Excretion of virus in the faeces has been detected as soon as the first day of pyrexia (CURASSON, 1932; 1942), on the 2nd to 7th day (HALL, 1933) or not earlier than the 3rd day (LIESS and PLOWRIGHT, 1964). It may occur in the absence of diarrhoea but possibly persists longer in the faeces of animals with signs of prolonged enteritis (HORNBY, 1926b; CURASSON, 1932; HALL, 1933). The faecal excretor rate in cattle infected with the RGK/1 strain rose to a maximum of 40% on the 7th day of the disease but virus had disappeared on the 9th day, even in animals with persistent diarrhoea. The maximal faecal titres varied between ca. $10^4 - 10^6$ TCD_{50}/g (LIESS and PLOWRIGHT, 1964) and faecal excretion must be considered one of the main methods of dissemination of the virus.

(iv) Other Excretions

The conjunctival and oral secretions are undoubtedly infective but no details are available. Vaginal exudates have been reported to be infective, even at intervals of 5—12 weeks after infection in those females which abort (CURASSON, 1921; JACOTOT, 1931c). Milk also contains virus which has been said to persist in recovered animals for up to 45 days (CURASSON, 1942).

G. Variation
1. Immunological Homogeneity of Rinderpest Virus Strains

It is universally agreed that all strains of rinderpest virus are immunologically very similar, if not identical. Perhaps the most convincing evidence of this lies in the fact that live vaccines such as the caprinised, lapinised (NAKAMURA III) and lapinised-avianised (LA) strains have been employed successfully in many countries of Africa or Asia where the disease is enzootic; they produce an immunity which is completely effective against natural or experimental challenge with any virulent strain of rinderpest. In this characteristic rinderpest virus, of course, resembles those of measles and canine distemper.

JACOTOT (1931 d) obtained formal confirmation of reciprocal-immunity reactions in cattle infected with four Indochinese strains of virus and in this laboratory cattle recovered from infection with many field isolates have always survived parenteral challenge with large doses of the RBOK stock laboratory strain. Nevertheless, some evidence was obtained that the latter strain was more readily neutralised than recent isolates, either by homologous or heterologous antisera, produced in cattle. It was suggested that this might have been due to the greater avidity for the RBOK strain or else to greater quantities of non-infectious antigen in the culture preparations of field isolates (PLOWRIGHT, 1963a). DEBOER (1961) observed 4-fold to 8-fold differences in the neutralisation titre of standard antisera against 10^3 TCD$_{50}$ of 3 culture-propagated strains (RBOK, Pakchong and Pendik). NAKAMURA (1931) used hyperimmune sera against a stock Japanese virulent strain, "O", and a recent Korean isolate "K", in crossed complement-fixation tests; he was able to differentiate the two viruses serologically. Minor antigenic differences between strains may be worth re-investigating in the light of these findings.

The most important variations which occur in rinderpest virus isolates are in their host affinity, pathogenicity and transmissibility. These will be discussed in the following sections.

2. Variations in Host Range

It is usually difficult to determine the natural host range of a particular strain of rinderpest virus. Subclinical infections can occur in some species and hence serological examination is necessary for a proper assessment of infection rates; the degree of exposure may vary considerably, due to such factors as feeding and watering habits, or population size and structure; thirdly the virus may change its apparent host preference in the course of a prolonged outbreak. An example of the latter is provided by the great African panzootic of 1889—1902; according to LUGARD (1893) it was curious that the wildebeest escaped when many other species were almost wiped out; yet PERCIVAL (1918) later reported the Masai assertion that wildebeest began to die last, when the cattle, buffalo, eland, etc. had all been decimated. It is now established that wildebeest are definitely very susceptible to rinderpest infection, although the mortality may not be high in some outbreaks (CORNELL, 1934; PLOWRIGHT, 1963c; PLOWRIGHT and McCULLOCH, 1967).

The ability to spread naturally in sheep and goats is another property on which some evidence is available. Rinderpest in goats and sheep in India was originally rarely encountered (JACKSON and CABOT, 1930) but become frequent during and after the 1930s (D'COSTA and SINGH, 1933; BAWA, 1940; LALL, 1947; DHANDA and MANJREKA, 1952) perhaps, it must be admitted, as the result of using low-passage goat virus as a vaccine for cattle (CRAWFORD, 1947; DAUBNEY, 1949). No records of natural rinderpest in sheep or goats in Nigeria could be traced prior to 1957, when the virus suddenly spread from cattle at the laboratory to both these small ruminants, causing typical clinical signs and a mortality similar to those in cattle. In E. Africa the virus has been known to affect sheep without spreading to goats (SCOTT, 1955b) and goats without apparently involving sheep (LIBEAU and SCOTT, 1960). In India the disease in sheep and goats did not always spread to cattle (BAWA, 1940; LALL, 1947; SHARMA, 1965).

The most extreme form of species adaptation yet discovered is that of the virus of "peste des petits ruminants" (GARGADENNEC and LALANNE, 1942; MORNET et al., 1956a, b). This virus does not apparently spread at all by contact from sheep and goats to cattle, although on inoculation to the latter it confers a solid immunity without producing noticeable clinical reaction. A similar virus causing asymptomatic infection of sheep and goats in Nigeria has been postulated by ZWART and ROWE (1966) as a result of a survey for rinderpest antibodies.

3. Pathogenicity

It is not easy to make an objective assessment of the pathogenicity of rinderpest virus for a particular species of domestic animal, because of the considerable genetic variation in susceptibility within the species. However, it is recognised that some isolates, such as the Pendik (Turkish) laboratory strain, have an extremely high virulence for cattle. NICOLLE and ADIL BEY (1899) also noted that they had failed to observe any recovery in cattle of the highly susceptible breeds (or crosses with these) which were inoculated with the strain of virus current at that time in Turkey. During the panzootic in South Africa, HUTCHEON (1902) noted that the usual recovery rate was about 10% but rose to 15—30% in some outbreaks. In two outbreaks in W. Africa (Cameroons and Congo) in 1960 and 1961 the cattle had long been free of the disease and the morbidity was nearly 100% with a mortality of ca. 85% (PROVOST and BORREDON, 1963).

At the present time in E. Africa naturally-occurring strains appear to fall into 2 main types, so far as their pathogenicity for cattle is concerned. Those isolated in S. Kenya and N. Tanzania in the past 10 years or so have invariably produced mild disease in grade (cross-bred zebu-exotic) animals, with a very low mortality in uncomplicated cases (ROBSON et al., 1959; PLOWRIGHT, 1963a; TAYLOR, 1966 — unpublished). These strains produced typical mouth lesions in about 70% of cattle and diarrhoea in only about 20% but spread readily by contact; 8—10 serial passages in cattle did not cause the virus to become lethal but there may have been some increase in the severity of the clinical signs (ROBSON et al., 1959; PLOWRIGHT, 1963a). Such strains of virus were probably current in this area during the previous 10 years at least, since a strain with the same low and stable pathogenicity was described by LOWE et al. (1947). Strains of similar low virulence have been isolated frequently in recent years in parts of W. Africa (PROVOST and BORREDON, 1963).

In the north of the E. African territories, however, the strains of virus isolated during the last few years have been of higher cattle pathogenicity and kill very large numbers of game animals. Hence the RGK/1 strain, isolated from a reticulated giraffe in this area during 1962, caused 48—84% mortality in grade cattle in different experiments (LIESS and PLOWRIGHT, 1964). This strain was probably similar to one referred to by MACOWAN (1961) as killing many cattle at the laboratory and causing widespread mortality in wild ungulates (STEWART, 1964). Further isolates of relatively high pathogenicity were obtained in 1965—1967 (TAYLOR, 1966; 1967 unpublished).

It is an attractive hypothesis that prolonged maintenance in game animals, as probably occurred in the southern sector (PLOWRIGHT, 1963c) was associated with the selection of virus attenuated for cattle, whereas in the northern sector this had not occurred as evidenced by the fact that at first game animals of all ages were affected (STEWART, 1964). This kind of interpretation has not been accepted by others, e.g. by RECEVEUR (1954) who did not regard game animals, by themselves, as potential reservoirs of virus.

Strains passaged in experimental hosts appear after variable intervals to reach a fixed though different pathogenicity for cattle and other domestic species, whilst increasing their virulence for the experimental animal [for details see section F 2(b) on experimental hosts]. These "fixed" attenuated strains are remarkably stable during many direct passages in cattle, which is extremely fortunate when their extensive use as live vaccines is considered. Details for caprinised virus are given by WADDINGTON (1945); SCOTT and WITCOMB (1958b — 50 passages) and ANON (1966a — 90 passages); for lapinised virus by NAKAMURA et al. (1943); IYER and SRINAVASAN (1954) and SCOTT and WITCOMB (1958b — 17 passages) and for culture-adapted strains by PLOWRIGHT, 1962c — 7 passages) and JOHNSON (1962c — 5 passages); the latter authors used virus of 90 and 65 culture passages, respectively.

On the other hand HALE and WALKER (1946) found that the Grosse Isle egg-modified strain reverted to virulence and re-acquired the capacity to spread by contact after 6 direct cattle passages; similarly, PLOWRIGHT and FERRIS (1962b) found that after only 41 culture passages the RBOK strain re-acquired some pathogenicity, within a single cattle passage, but it did not spread by contact after 4 cattle transfers.

4. Transmissibility

The ease with which strains of rinderpest spread naturally varies considerably but little is known of underlying differences in the viruses. It is a mistake to regard the majority of strains of rinderpest virus as transmissible quickly and consistently by natural routes even when the donor animals are in their most infectious state — i.e. in the middle or late pyrexial phases [see Section F 3(e) on routes of excretion]. COOPER (1932), for example, carried out experiments at Mukteswar which showed that strains of virus causing about 60—70% mortality failed to infect some cattle by stall contact after 9—10 days exposure; IDNANI (1944) found that a highly virulent strain, causing >90% mortality, spread readily by expired air over a distance of 6 feet but another strain, causing about 12% mortality, would not transmit at all over this distance. The stock laboratory strain RBOK, now passaged by subcutaneous inoculation for over 55 years and producing 75% mortality (MacOWAN, 1956) does not spread at all readily by contact, neither does it produce any characteristic mouth lesions (BROTHERSTON, 1951b; PLOWRIGHT, 1952). In early culture passages, however, this virus rapidly re-acquired both the capacity to spread by contact and to cause extensive mouth lesions (PLOWRIGHT and FERRIS, 1959b). These two characteristics could be associated.

Attenuated strains which have been employed as live vaccines never spread by contact. Thus PFAFF (1940), WADDINGTON (1945) and DAUBNEY (1947) failed to observe spread of goat-adapted virus in cattle, although it has been demonstrated in the excretions of infected animals (CORNELL and OONGWONGSE, 1941; MITCHELL and LE ROUX, 1946). Lapinised and avianised strains are also incapable of spreading by contact (BROTHERSTON, 1951a; JENKINS and SHOPE, 1946). The inability of culture-attenuated virus to extent from inoculated cattle to others in close contact is probably associated with its failure to proliferate in the alimentary and respiratory mucosae or in the parenchymatous organs which could contribute virus to the excretions (TAYLOR and PLOWRIGHT, 1965).

H. Immunity
1. Active Immunity

Infection with virulent strains of rinderpest virus is often considered to confer a solid and lifelong immunity to parenteral or natural challenge. CURASSON (1942) concluded, however, that the immunity which followed a natural or experimental infection lasted always for several years, *not* usually for life, and that the exact duration depended on the severity of the original attack and on natural reinfections; he also noted that PIOT BEY (1930) found that cattle were consistently immune up to 7 years after sero-infection.

Probably the best evidence on the duration of immunity is that provided by cattle inoculated with goat-adapted virus and subsequently kept in a rinderpest-free environment, on the island of Pemba; BROWN and RASCHID (1965) found that 17 such cattle failed to react clinically to challenge inoculation at twelve and a quarter to twelve and a half years after vaccination; all these animals possessed neutralising antibody in their serum before challenge and one of four showed an anamnestic response. Three other cattle said to have been immunised were in fact serologically negative, reacted to challenge and did not show an anamnestic response. DATTA (1954) also reported that caprinised virus had been shown to confer a solid immunity which lasted at least 14 years in Indian cattle. The report of BROWN and RASCHID (1965) is distinguished from many older publications, which purport to describe the waning of immunity following live-virus vaccination, in that antibody was sought before the challenge procedure and that an attempt was made to differentiate anamnestic from primary serological responses.

Cattle infected with avirulent, culture virus of the 40th BK passage retained antibody to the end of the fourth year and were immune to parenteral challenge with virulent virus; serological responses occurred in animals in which serum antibody had fallen below a $\log_{10} SN_{50}$ titre of 1.2 (PLOWRIGHT, 1962c). Anamnestic responses were also seen in some animals which had been immunised with lapinised virus 15 months previously and were then given a formalin-inactivated vaccine (PLOWRIGHT, 1962b). Cattle given virus of 91 culture (BK) passages remained clinically resistant to parenteral challenge for at least 6 years but some grade animals with very low levels of antibody showed striking and rapid serological responses even after as short a time as 27—37 months; in 2 or 3 cases it was possible to show that the challenge virus proliferated in them (PLOWRIGHT and TAYLOR, 1967 and unpublished). The duration of immunity following inoculation with lapinised virus was at least 49 months (BIRKETT, 1958), or 7 years according to MENON and SAGAR (1963), but no antibody studies accompanied these observations.

It is difficult to avoid the impression that, in those cases where serological tests have been used to control the immediate efficacy of any live virus vaccine, immunised cattle do not become susceptible to challenge again, no matter how long they are held. The apparent failures to confer long-term immunity could probably have been discovered in the short-term — 2—3 weeks after vaccination; it has yet to be proved that any animal has been successfully immunised with a live rinderpest vaccine and has later become clinically susceptible to parenteral or natural challenge. The same proposition has been made with respect to canine distemper (PLOWRIGHT, 1964c).

The opposite point of view has been put succinctly by SCOTT (1964) who formulated the following hypothesis, viz: "there is a direct relationship between the duration of immunity and the peak of the virus growth curve which, in turn, governs the clinical response". If we accept the importance of neutralising antibody as an indicator of immunity, the dual fallacies which lie behind this hypothesis are (a) that the neutralising antibody response to strains of rinderpest virus is determined quantitatively and directly by their pathogenicity; and (b) that there is a predictable decline in the titre of antibody with time and this eventually entails a loss of resistance to clinical reinfection.

The first part of the hypothesis is patently mistaken, since there is no evidence that the immediate neutralising antibody response to fully virulent strains, e.g. RBOK, RGK/1 (PLOWRIGHT, 1962b) or recent mild field isolates (PLOWRIGHT, 1963a) differs significantly from that to caprinised, lapinised or culture-attenuated vaccines (PLOWRIGHT, 1962b; 1962c; PLOWRIGHT and TAYLOR, 1967). The same conclusion with respect to eight virulent and attenuated strains was reached by SCOTT and BROWN (1958), using rabbits to measure antibody, as opposed to the culture system employed by later workers. With respect to the second deduction — that there is a predictable decline in the titre of antibody with time, leading eventually to a return to clinical susceptibility — there is again no supporting evidence and much to the contrary. Thus, the mean titre of neutralising antibody produced in response to apathogenic, culture-attenuated virus declined in grade cattle over the first 18 months and then remained virtually constant to four and a half years; in short-horn zebus and Ankole long-horn cattle it hardly fell at all over periods of 4 and 2 years respectively. Individual animals showed considerable variations, some exhibiting a marked decline ($\gtrsim$ 10-fold), others none at all; but, even when circulating antibody was no longer detectable, individual animals were still completely resistant to challenge, at least from a clinical point of view. (PLOWRIGHT and TAYLOR, 1967.) If this is accepted for a virus strain which is absolutely non-pathogenic for grade cattle then the hypothesis that duration of immunity is directly related to the severity of the clinical response must be abandoned.

2. Neutralizing Antibody

According to CURASSON (1932, 1942) the first demonstration of the protective effect of the serum or milk of recovered animals was provided by SEMMER (1893) but it was not until 1940 that an *in vitro* neutralisation test became available, utilising rabbits as test animals and the NAKAMURA III strain of lapinised virus (NAKAMURA, 1940). Titres were not affected by heating sera for 30 minutes at 56°C and the subcutaneous route of inoculation gave higher titres than the intravenous (NAKAMURA, 1940). SCOTT and BROWN (1958) used an essentially similar system, mixing 5- or 10-fold dilutions of inactivated serum with $10^{1.3}$ to $10^{2.3}$ ID_{50} of virus and holding the mixtures either for 1 hour at 37°C or 20 hours at 4—10°C; each mixture was then inoculated into 5 rabbits intravenously and all were killed after 5 days to detect the typical lesions of lapinised rinderpest. The conditions of incubation of the virus-serum mixtures did not influence the serum titre, a finding which confirmed the results of NAKAMURA (1940); the dose of virus was, however, very important since, from data provided by SCOTT and BROWN (1958); it can be calculated that a 10-fold increase in virus might, on average, lower the serum titre by 1.0 to 1.6 log units.

HUARD et al. (1959) advocated a constant serum:virus dilution method, requiring 8 to 16 rabbits per sample compared with the 21 to 35 rabbits used by SCOTT and BROWN (1958); this type of test has continued to be used by some American workers (e.g. STONE and DELAY, 1961; DELAY et al., 1965), in spite of its costliness and relative inaccuracy. JENKINS and WALKER (1946) used a rabbit-adapted strain of virus which produced only a fever in inoculated rabbits; they mixed 2 dilutions of a bovine spleen suspension containing the virus with undiluted cattle sera and found that neutralisation occurred only with the serum of calves which had been immunised with inactivated or chick-embryo-attenuated virus. They therefore concluded that "there is a definite correlation between immunity and the development of neutralising antibodies".

WALKER et al. (1946a) demonstrated neutralising antibodies by mixing undiluted serum with 10^3 or 10^4 cattle ID_{50} of virus in the form of a cattle spleen suspension. They held the mixtures for 2 hours at 34°F before inoculating them into calves, which were protected by immune serum mixtures.

A major incentive for the development of tissue culture techniques for rinderpest virus was to produce economical methods for detecting and titrating rinderpest-neutralising antibody (PLOWRIGHT and FERRIS, 1959a). These techniques were rapidly developed after the demonstration of easily recognised cytopathic effects for the virus and, in E. Africa, have been almost entirely of the constant virus:serum dilution type, whereas JOHNSON (1962b) employed the constant serum:virus dilution technique. PLOWRIGHT and FERRIS (1961c) used unheated cattle sera but inactivated serum from all other animals at 56°C for half-an-hour. JOHNSON (1962b) heated all sera at 56°C for 1 hour, primarily to reduce microbial contamination; careful comparative tests in this laboratory showed no significant difference in the titre of standard sera tested before and after heating at 56°C for half-an-hour (PLOWRIGHT and HERNIMAN — unpublished). On the other hand, human sera contained rinderpest-neutralising antibody, the titre of which was often lowered considerably by heating and restored by the addition of guinea-pig complement; this antibody was, however, a heterologous one, appearing in response to measles infection (PLOWRIGHT, 1962a).

The time and temperature of incubation of virus:serum mixtures was about 18 hours at 4°C (PLOWRIGHT and FERRIS, 1961c; PLOWRIGHT, 1962b) or 1 hour at 37°C (JOHNSON, 1962b). A recent investigation of the best conditions for neutralisation of rinderpest virus by cattle antisera showed that the majority of the reaction occurred almost immediately either at 4°C or 37°C, but the virus-antibody complexes were still relatively unstable after as long as 4 hours at 37°C or 24 hours at 4°C, so that some reactivation took place on dilution. The most stable complexes were probably formed after 1 hour at 37°C, followed by 18—24 hours at 4°C in PBS, pH 7.2 and containing 0.1% bovine plasma albumin (PLOW-RIGHT and HERNIMAN — unpublished).

The relationship between the dose of virus employed and the titre of the serum in the tissue culture system was examined by PLOWRIGHT and FERRIS (1961c). They found that a 1-$\log_{10}$ increase in the test dose of virus could be expected to result in a decrease of about 0.5 to 0.6 $\log_{10}$ units in the SN_{50} titre of a serum. It was also calculated that titre differences of 0.6 $\log_{10}$ units or greater were probably significant in any one series of tests. Between tests there was a somewhat greater variation, since the mean and standard deviation of titres calculated for a reference antiserum, diluted 10-fold, was $10^{2.24\pm0.29}$ in primary BK cells; in serially-cultivated BK cells the figures for another reference serum were $10^{2.66\pm0.14}$ (PLOWRIGHT and FERRIS, 1961c). The permissible range of virus dose was regarded as $10^{1.8}$ to $10^{2.8}$ TCD_{50} per culture tube.

The majority of cattle convalescent from infection with virulent or attenuated strains of virus show serum titres in the tissue culture system ranging from $10^{1.5}$ to $10^{3.5}$ but a few animals never develop titres as high as $10^{1.0}$ (PLOWRIGHT, 1962c; PLOWRIGHT and TAYLOR, 1967). In the rabbit test as described by SCOTT and BROWN (1958) the geometric mean titre and 95% fiducial limits for all cattle in which strains of rinderpest virus multiplied was $10^{3.1\pm0.3}$; nevertheless, neutralisation at a titre of $10^{1.0}$ was necessary to indicate immunity, whereas any complete neutralisation with undiluted serum in the culture system was a reliable guide to resistance to challenge (PLOWRIGHT, 1962b). Screening tests, for determining the susceptibility of experimental cattle to rinderpest, have been carried out in both rabbits (BROWN and SCOTT, 1959) and tissue cultures (PLOWRIGHT, 1962b; JOHNSON, 1962b).

JOHNSON (1962b) employed a constant serum:virus dilution technique for quantitative culture tests in Nigeria, expressing serum titres as a neutralisation index (NI). Sera with an NI of 10^4 were regarded as "strong"; others, with an NI of 10^2 as "weak". SMITH (1966) later found that calves with passively-acquired antibody, giving an NI of $10^{2.1}$ or less were susceptible to the inoculation of rinderpest culture vaccine; the minimal NI which could be detected was $10^{1.7}$ and the critical level above which calves failed to respond to vaccination was given by an NI of $10^{2.3}$ to $10^{2.0}$.

Japanese workers have employed avianised strains of virus for neutralization tests in embryonated chicken eggs (FURUTANI et al., 1954; NAKAMURA et al., 1955; NAKAMURA, 1957b) but unfortunately the results could only be read indirectly, after performing complement-fixation tests for viral antigens on pools of chick embryo spleens. After the LA strain of virus had been adapted to culture in fragments of chick embryo this host system was also employed, again in conjunction with complement-fixation tests, to detect and titrate neutralising antibody (NAKAMURA et al., 1958; NAKAMURA, 1957b). Following recognition of a high, specific mortality in eggs inoculated with the BA strain of virus, PIERCY et al. (1958) claimed success in detecting neutralising activity in sera by the suppression of embryo mortality; this test was never apparently developed further, although it was claimed to give good quantitative agreement with the results obtained in rabbits. The most recent development in techniques for estimating neutralising antibody is the use of newborn mice and a mouse neurotropic strain of rinderpest virus (IMAGAWA, 1965). Antibody titres in this system were about 2—4-fold lower than those estimated simultaneously in cell cultures.

The time of appearance of neutralising antibody in cattle was found to be the 5th or 6th days post-infection in animals inoculated with large doses of the virulent RBOK strain (SCOTT and BROWN, 1958; PLOWRIGHT, 1962b) and the 4th day of pyrexia in cattle reacting to the RGK/1 strain (PLOWRIGHT and LIESS — unpublished). Culture-attenuated virus stimulated antibody formation on the 6th day after inoculation of $10^{4.5}$ TCD_{50} (PLOWRIGHT and FERRIS, 1959b) or from the 7th to 17th days with decreasing virus doses (JOHNSON, 1962c). Some cattle given small doses of culture-attenuated rinderpest virus did not develop antibody by the end of the 2nd week (PLOWRIGHT and FERRIS, 1959b) or even after 3 weeks and yet they were immune to challenge (PLOWRIGHT, 1962c); a similar finding has been recorded for caprinised virus (JOHNSON, 1962b). The explanation for these observations is probably that antibody does not become detectable until the 4th week or later in some animals given minimal immunising doses of these vaccines, and not that a permanent dose-response relationship is involved (cf. PLOWRIGHT, 1962b).

In Nigeria a small proportion of non-humped cattle (ca. 0.25%) did not develop antibody after 3 months (JOHNSON, 1962b). PROVOST et al. (1965a) showed that some animals did not develop antibody even after challenge and they concluded that this was due to an essential hypoglobulinaemia, since antibodies against other cattle viruses were also absent and γ-globulin levels were very low.

Peak levels of neutralising antibody were usually reached between 2 and 4 weeks after infection (PLOWRIGHT, 1962b, c; JOHNSON, 1962b, c; PLOWRIGHT and LIESS — unpublished). A small decline in mean titre was detected in some groups of animals after 6 months (PLOWRIGHT, 1962b, c; PLOWRIGHT and TAYLOR, 1967) or 8 months (JOHNSON, 1962b) and this fall sometimes increased to 18 months but not invariably. Some individuals showed no detectable decline of serum titre over periods of 2 to 4 years, whereas in others titres declined rapidly to barely detectable levels (PLOWRIGHT and TAYLOR, 1967).

3. Complement-fixing Antibody

Complement-fixing antibodies were detected over 30 years ago in the sera of cattle convalescent from rinderpest and were shown to be dependent on a thermolabile factor destroyed by heating at 55°C for 30 minutes (NAKAMURA, 1958). The Japanese workers compared the amount of complement fixed by mixtures of fresh, unheated serum and antigens derived either from normal or from infected bovine lymph nodes; the latter were extracted after drying over calcium chloride at room temperature and fixation occurred over 4 hours at 4—7°C. Details of the Japanese techniques were given by NAKAMURA (1958) and NAKAMURA and MACLEOD (1959). WALKER et al. (1946a) attempted to use calf spleen antigen and heated sera in complement fixation tests but found that the titres obtained were too low to be dependable. COOPER (1946) used allantoic and amniotic fluids, from infected chick embryos, as antigen and found that 91/95 immunised calves developed heat-stable antibody titres of 1/6 to 1/500 by the 13th—14th days after vaccination with egg-attenuated virus; a fixation period of 18—22 hours at 4—7°C was employed. PELLEGRINI and GUARINI (1952) used an antigen prepared from the lymph nodes and buccal and gastric mucosae of cattle and diluted sera in 1.5% sodium chloride to destroy the non-specific activity; they obtained titres of 1/30—1/40 in immune animals. MOULTON and STONE (1961) used cattle lymph node extracts as antigen and 18 hours fixation at 4°C; the sera were heat-inactivated and the amount of complement fixed was estimated.

Complement-fixing antibodies appear in the serum of cattle on the 9th to 17th days (COOPER, 1946), between 1 and 2 weeks (NAKAMURA, 1958) and by 10 days (MOULTON and STONE, 1961) after inoculation. They reach a peak in 14 to 18 days and remain demonstrable in some animals for as long as 6 months (COOPER, 1946) but may disappear within a few days (NAKAMURA, 1958). In rabbits complement-fixing antibodies were demonstrable between 7 to 10 days, reached a peak at 15 to 20 days and persisted in some for over 70 days (NAKA-MURA, 1958). The level of antibody depended, according to NAKAMURA (1958), on the virulence of the infecting virus. The development of complement-fixing antibodies is not a reliable guide to the efficacy of immunisation procedures; for example, only 57 (62%) of 91 Korean cattle became serologically-positive after inoculation with LA virus vaccine but 25 of the negatives or doubtfuls did develop neutralising antibody (NAKAMURA, 1957a). The short-lived character of complement-fixing antibodies also greatly reduces their value in epizootiological work or in studies on the duration of immunity following vaccination.

4. Precipitating Antibody

Antisera for use in agar-gel diffusion tests are prepared by the hyperimmunisation of rabbits (WHITE, 1958b; SCOTT and BROWN, 1961), cattle (SCOTT, 1962a; ISHII et al., 1964) or goats (SCOTT et al., 1963). Precipitating antibodies cannot, however, be demonstrated in the serum of convalescent animals (WHITE, 1958b; THOME, 1965) although unsubstantiated claims have been made to this effect (HUSSAIN and SARWAR, 1962). An indirect gel-diffusion test, depending on the prior neutralisation of a stock precipitinogen, was used by WHITE and SCOTT (1960) for the demonstration of antibodies in convalescent sera; it has not been developed further.

5. Antibody Inhibiting Measles Haemagglutinin

It was shown by WATERSON et al. (1963) that during the course of virulent rinderpest infection in cattle, antibodies appeared which inhibited measles haem-agglutinin to high titre. This finding was utilised by BÖGEL et al. (1964) for the demonstration of antibodies in cattle convalescent from infection with virulent or attenuated strains of rinderpest virus. These antibodies appeared 9—12 days after virulent infection and reached maximal titres of 1/32 to 1/64 (BÖGEL et al., 1964). The titre depended, as with complement-fixing antibodies, on the virulence of the infecting virus (THOME, 1965) and remained stable for at least several months, though later it declined to non-detectable levels in some animals; on this account the test was not recommended for serological surveys of rinderpest immunity (ANON, 1966b).

6. Passive Immunity

Antisera produced by the hyperimmunisation of cattle or derived from convalescent animals were used for very many years in the simultaneous serum: virus method of immunisation. The method was used extensively in S. Africa from 1897/98, where it had been developed by TURNER and KOLLE, following attempts by DANYSZ and BORDET to immunise cattle by exposing them to contact infection after administration of serum (HUTCHEON, 1902). Antiserum was also used alone to protect animals involved in outbreaks of the disease but its use for this purpose has virtually ceased since the discovery of protection by "interference", within a short period after the inoculation of live attenuated vaccines.

The most important aspect of passive immunisation is the effect that colostral antibodies have in protecting the young of immune dams in enzootic areas and in preventing their active immunisation by live vaccines. The literature has been reviewed by BROWN (1958a) who investigated the problem in cattle by use of a neutralisation test carried out in rabbits (BROWN, 1958b, c). No antibodies were found in the sera of calves prior to suckling but they were present 30—48 hours after the ingestion of colostrum; the antibody titre in colostrum samples was greater than that in the calves, which in turn had higher serum titres than their dams. This passively-acquired antibody declined exponentially with time, its half life being 37 days and the extinction point 10.9 months.

BROWN (1958c) found that all calves which possessed neutralising antibody to a titre of $10^{0.7}$ or less, were susceptible to the inoculation of live caprinised virus and responded actively, whereas all those with titres $\leq 10^{2.2}$ failed to become infected or sensitised. Calves with serum titres between $10^{0.7}$ and $10^{2.2}$ could sometimes be infected but in other cases were not; in this group the amount of antibody produced within 3 weeks by animals actively immunised was inversely proportional to the titre of passively-acquired antibody at the time of inoculation; antibody titres at 1 year after inoculation showed a similar, though less marked, inverse relationship. A comparable connection between trace amounts of passively-acquired antibody and the degree of serological response in 8—11 months old calves inoculated with culture-attenuated vaccine was postulated by PLOWRIGHT (1962c).

STRICKLAND (1962), basing his figures on a pyrexial response to caprinised virus, found that 50—65% of calves from immune dams were susceptible by the age of 4—6 months; 90% by 7—8 months and 95% by 9—10 months. BROWN (1958c) and SMITH (1966) obtained essentially similar results employing serological techniques, but BROWN (1958c) found that some calves reacting serologically showed no clinical response to caprinised virus. SMITH (1966) used neutralisation indices estimated in a tissue culture system, and found that the extinction point for passively-acquired antibody in Nigerian calves was 10.9 months; he also found that the level of maternal antibody had fallen virtually to zero before the calf was susceptible to the inoculation of rinderpest culture vaccine.

In a population of free-living wildebeest *(Connochaetes taurinus)* PLOW-

RIGHT and McCULLOCH (1967) found that rinderpest antibody titres in the calves normally exceed those of their dams during the first 6 weeks of life and individual calves became negative from the 10th week onwards. The half-life of passively-acquired antibody was 4.4 weeks and the estimated extinction point was ca. 7 months. It is probable that other highly susceptible game species, such as eland or buffalo, behave in a similar manner in enzootic areas.

7. Interference

All those attenuated strains of rinderpest virus which are used as live vaccines confer protection before the development of antibody, the time to establishment of the resistance, depending, probably, on the time required for generalisation of the attenuated strain, which in turn is dependent on the dose administered. Thus PFAFF (1938) found that cattle and buffaloes resisted challenge with virulent virus only 48 hours after the inoculation of caprinised virus; when cattle received about 20—40 MID of caprinised virus this figure was confirmed by WILDE and SCOTT (1961), who observed no protection after 33 hours. The Nakamura III strain of lapinised rinderpest virus protected in 84—108 hours (BROTHERSTON, 1951a; ILLARTEIN and GUERRET, 1954; SIMPSON, 1954) and the Grosse Isle strain of avianised virus in 96 hours, although antibody did not appear until the 8th day (HALE et al., 1946).

Using culture-attenuated virus, PLOWRIGHT and FERRIS (1959b) found that a dose of $10^{4.5}$ TCD_{50} gave protection to one of two grade cattle within 72 hours and to both of two within 96 hours; the absence of circulating antibody was confirmed at the latter time. With only $10^{2.5}$ TCD_{50} of the attenuated strain resistance was not established until the 5th day. The dose of virulent virus for challenge inoculation in all these experiments was probably ca. 10^5 to 10^6 cattle ID_{50}.

JOHNSON (1962c) found that $10^{1.7}$ TCD_{50} of culture-attenuated virus protected 2 cattle against caprinised virus infection at 96 hours but that antibody was not detectable until 10—12 days had elapsed. TAYLOR and PLOWRIGHT (1965) showed that rinderpest culture vaccine in a dose of $10^{4.0}$ to $10^{4.6}$ TCD_{50} had generalised within 96 hours of subcutaneous inoculation but not at 72 hours; antibody did not appear until the 7th day at the earliest. There is thus good agreement between the time of onset of resistance and the time for widespread dissemination of culture-attenuated virus.

SCOTT and WITCOMB (1958a) recorded an interesting example of heterologous interference by rinderpest virus in hamsters. When these animals had been infected with rinderpest 2—7 days previously, they resisted inoculation with Rift Valley fever virus; nevertheless they remained normally susceptible to later challenge with this rapidly lethal agent.

8. The Immunological Relationships between Rinderpest and Other Viruses

It has already been noted briefly that rinderpest has morphological and serological relationships with the viruses of human measles and canine distemper (see Classification and Nomenclature). Research on the immunological relationships will now be more fully discussed.

a) Rinderpest and Measles

The effect of rinderpest virus in man is unknown but rinderpest neutralising antibody was present in the serum of nearly all human adults (PLOWRIGHT and FERRIS, 1959a; WARREN, 1960), and appeared in children or monkeys during convalescence from measles (PLOWRIGHT, 1962a; DELAY et al., 1965). The titre against the heterologous virus was probably lower than against the homologous one and appeared, in some cases at least, to be complement dependent (PLOWRIGHT, 1962a). Monkeys inoculated with rinderpest virus sometimes showed mild febrile reactions and usually developed homologous neutralising antibody to very low titre but it was not possible to demonstrate protection against challenge with measles virus, as control monkeys did not show marked clinical signs (DELAY et al., 1965). The sera of rinderpest-immune cattle contained antibodies to measles virus (IMAGAWA et al., 1960) and, in one system, these had a higher titre than those against the homologous virus (PLOWRIGHT, 1962a). A single dose of virulent measles virus ($10^{3.6}$ or $10^{6.4}$ TCD$_{50}$) failed to protect cattle against challenge with rinderpest virus (PLOWRIGHT, 1962a; DELAY et al., 1965) and no antibody was produced to either agent; these animals are probably, therefore, not capable of supporting the growth of measles virus.

Rabbits were not protected against challenge with lapinised rinderpest virus by a single dose of measles virus; three or more inoculations by various routes, with and without adjuvants, caused the production of rinderpest-neutralising antibody and conferred a complete resistance to challenge with lapinised rinderpest virus (PLOWRIGHT, 1962a). It can probably be concluded that rabbits were not susceptible to the strain of measles virus employed but could be immunised against the heterologous virus by a sufficient large antigenic mass.

Attempts to demonstrate a common antigen of rinderpest and measles viruses, using hyperimmune serum and cell-culture antigens, failed completely (PLOWRIGHT, 1962a), but LIESS (1963) was able to show specific immunofluorescence in measles-infected cells by use of rinderpest-immune globulin prepared from hyperimmune cattle serum.

b) Rinderpest and Canine Distemper

Since the original observations of POLDING and SIMPSON (1957) quite a voluminous literature has appeared on the immunological relationship between these two viruses, much of it apparently stimulated by the prospect of using distemper virus to immunise cattle against rinderpest; the majority of the literature has been summarised by MORNET et al. (1960).

Dogs inoculated with virulent (POLDING and SIMPSON, 1957; SCOTT and BROWN, 1958; POLDING et al., 1959; DELAY et al., 1965) or lapinised (MORNET et al., 1960) strains of rinderpest virus develop a viraemia and antibodies to the homologous virus but not to distemper.

Ferrets inoculated with virulent or lapinised rinderpest virus possibly produce homologous neutralising antibody (MORNET et al., 1960). In one experiment virulent Senegal virus protected 5 of 6 ferrets against challenge 26 days later with distemper virus (GORET et al., 1958; MORNET et al., 1959a) whilst no protection was conferred in another experiment (MORNET et al., 1960). Lapinised rinderpest virus (Nakamura III strain) protected a large proportion (14/18) of inoculated ferrets against challenge with virulent distemper virus, but it was not clearly shown whether the protection was dose-dependent (GORET et al., 1958; MORNET et al., 1959a). In subsequent experiments the resistance was found to be irregular from the 5th to 9th days after inoculation of the lapinised virus but thereafter it was invariably present; it lasted at least eleven and a half months, i.e. it was not due to "interference" (GORET et al., 1960b).

Rabbits which received large quantities of virulent or avianised canine distemper virus were not protected against challenge with lapinised rinderpest virus (GORET et al., 1960a), no matter what the route of inoculation employed or if it was repeated 2—4 times without or with adjuvants (MORNET et al., 1960; PLOWRIGHT, 1962a).

Cattle inoculated with virulent or avianised distemper virus, in the form of ferret spleen and brain or egg membranes, respectively, did not show clinical reactions but were protected against rinderpest challenge (GORET et al., 1958; MORNET et al., 1960). POLDING et al. (1959) failed to observe resistance in a similar experiment, even when giving repeated doses of distemper virus of unspecified type. The dose of canine distemper virus needs to be large to confer protection; thus, while 900 mg of lyophilised ferret spleen protected 4/4 cattle, only 2/4 were protected by 150 or 15 mg; 1 g or 10 g of infected egg membranes only protected half the inoculated cattle (MORNET et al., 1959a, b). Protection was accurately correlated with the production of antibodies to distemper virus (GILBERT et al., 1960) and hence, one can assume that the infection failed to become established in those animals which were not protected against rinderpest. DELAY et al. (1965) found, however, that virulent distemper virus (strain Synder Hill) did not protect cattle against challenge at 24 days, although 5 of 6 had distemper antibodies at that time.

The resistance to rinderpest conferred by distemper virus becomes demonstrable in some cattle by the 6th day but is not constantly present until after 12—18 days (MORNET et al., 1960); the duration of protection has not been determined but it lasts at least six and a half months (GILBERT et al., 1960). It has not been determined whether or not rinderpest virus multiplies in cattle which successfully resist challenge.

Rinderpest and distemper viruses certainly possess closely-related antigens which have been demonstrated by the Ouchterlony technique (WHITE et al., 1961). The antigens are not absolutely the same, however; a line of identity appeared when both antigens were diffused against hyperimmune distemper serum, prepared in a dog but spur formation resulted when rabbit hyperimmune rinderpest serum was employed.

Rinderpest antisera derived from convalescent cattle or prepared by the hyperimmunisation of cattle neutralised avianised (GORET et al., 1959; MORNET et al., 1960) or ferret distemper virus *in vitro* (GORET et al., 1959). Rinderpest hyperimmune serum prepared in rabbits, however, had no neutralising effect on avianised distemper virus (MORNET et al., 1960). Rinderpest antisera confer no passive protection when administered simultaneously with virulent distemper virus, either to dogs (POLDING et al., 1959) or ferrets (GORET et al., 1960c).

Distemper antisera prepared in dogs, horses, cattle and ferrets neutralise lapinised rinderpest virus but often to low titre (POLDING and SIMPSON, 1957; GORET et al., 1960a; MORNET et al., 1960; VILLEMOT et al., 1961). No neutralisation of bovine strains of rinderpest virus by hyperimmune distemper serum was demonstrable by MORNET et al. (1960) but VILLEMOT et al. (1961) had limited success.

J. Clinicopathological Features
1. Clinical Features

The clinical signs of rinderpest in cattle and other natural hosts are essentially the same but show wide variations in intensity and frequency, depending on the pathogenicity of the strain of virus involved and the resistance of the animal, whether conditioned by age, condition, breed or species. This variability was very well-known to those like CURASSON (1932, 1942) who had wide experience of the disease in enzootic areas, where strains of reduced virulence and cattle of lower sensitivity were commonly involved. It is not made so clear in some of the recent descriptions (e.g. MAURER et al., 1956) which are excellent so far as the acute, severe disease in highly susceptible cattle is concerned.

CURASSON (1932, 1942) described hyperacute, acute, subacute and chronic cases and divided the course of the typical, acute infection into 4 phases, i.e. incubation, invasion, external lesions and gastrointestinal. He recognised that one or other phase may be modified or fail to appear altogether and that cutaneous or nervous localisation occurred. As a sequel to a study of the pathogenesis of the disease in experimental cattle, PLOWRIGHT (1964a, 1965) suggested substitution of the term "prodromal" for the "invasion" phase and of "mucosal" for the "external lesions" and "gastrointestinal" phases of earlier authors. The justification for this was that virus "invasion" or generalisation had already occurred at the end of the incubation period and that proliferation proceeded simultaneously in both the externally visible mucosae — nasal, oral, probably also vaginal — and in those of the gastrointestinal tract [see Section F3 (c)]. It is only necessary here to indicate briefly the clinical signs observed during the prodromal and mucosal phases.

a) Prodromal Phase

This is characterised by fever of sudden onset, often reaching a peak of about 105—107°F (40.5—41.5°C) on the second or third days after onset. Accompanying pyrexia there is increasing depression or restlessness, some loss of appetite and a fall of milk yield in dairy cows. The visible mucosae become congested, the muzzle is dry and there are often slight serous or seromucoid discharges from the eyes and nostrils; these latter become more profuse and mucopurulent as the disease progresses. There is acceleration of the respiration and heart beat, suppression of rumination and constipation. This phase lasts on average about 3 days, the majority of animals beginning to show mucosal lesions between the 2nd and 5th days following onset of pyrexia. No specific or diagnostic clinical signs are visible during this prodromal phase.

b) Mucosal Phase

This commences with the appearance of foci of necrosis, superficial erosion and capillary haemorrhage in the mucosae of the mouth cavity, sometimes heralded 1—2 days earlier by similar lesions in the nasal, vulval or preputial mucosae; it is on the oral lesions, however, that diagnosis principally rests. The necrotic foci fuse to form more or less extensive and loosely-adherent deposits, present on the lips, gums and cheek papillae, the base, lateral and inferior aspects of the tongue, and the pharynx; circular plaques of necrotic papillae also occur on the dorsum of the tongue and excessive salivation is often observed.

During this time the general symptoms are exacerbated, the respiration becomes painful and laboured, often of an abdominal type and there may be an infrequent cough. Pneumonic complications are rare and due to secondary bacterial infection.

Diarrhoea appears usually on the 4th to 7th days of pyrexia, 1—2 days after the external lesions; it is at first watery, but later includes fresh or altered blood, excess mucus and even strands of necrotic intestinal mucosae; in severe cases it results in rapid dehydration, haemoconcentration, weakness, prostration and finally, a semi-comatose state.

Skin lesions, which have recently acquired an added interest in view of the taxonomic relationships of rinderpest to measles virus, appear during this phase especially on those parts of the body where the hair is fine, e.g. on the udder, inside the legs and around the mouth; they are seen more rarely on the dorsal aspects of the body. Skin lesions have been described in sheep and goats as well as cattle (MOHAN and BAHL, 1953; MORNET et al., 1956b). CURASSON (1942) described macules, succeeded by papules and vesico-pustules, which cause erection and tufting of the hair; more rarely there are larger areas of an eczematous exudation. There is little doubt that skin lesions are now seen less frequently than formerly and that they are often, but not invariably, associated with strains of low pathogenicity (CURASSON, 1942; MORNET and GUERRET, 1950). It is also of interest that many old vernacular names for rinderpest are the same as those used for small-pox (CURASSON, 1942; MACGREGOR, 1944). There is also little doubt that many cases of so-called "cutaneous" rinderpest were in reality caused by concurrent streptothricosis, a bacterial infection.

The "nervous form" of the disease, has never been proved to be due to rinderpest virus infection and no lesions of a meningitis or encephalomyelitis have been described in infected ruminants. It is possible that many cases of the "nervous" form have been due to accompanying protozoal or rickettsial pathogens (CURASSON, 1942; PROVOST and BORREDON, 1963).

Death usually occurs on the 6th to 12th days after onset of pyrexia, particularly during the later part of the mucosal phase; occasional deaths are delayed to the 3rd week. The mortality rate may be well over 90% in cattle of high susceptibility infected by virulent strains. Strains of low virulence produce a low or insignificant mortality in resistant breeds of cattle, at least when secondary infections are not present.

c) Convalescent Phase

Resolution of the visible mucosal lesions often becomes evident with dramatic speed about the 3rd to 5th days after their appearance; it may be complete in as little as 48 hours. Diarrhoea often persists for a longer time and complete recovery from a severe attack requires at least 4—5 weeks.

d) Mild Forms of Rinderpest

In these cases the general and mucosal lesions are of reduced degree and duration, but the most common feature in the author's experience is the mild and transient diarrhoea, or even the complete failure of enteric signs to develop (ROBSON et al., 1959; PLOWRIGHT, 1963a). Apyrexial forms are not infrequent (PLOWRIGHT, 1963a; PROVOST and BORREDON, 1963), whilst sometimes diarrhoea occurs without the development of mouth lesions (PROVOST and BORREDON, 1963).

2. Pathology

The gross pathology of bovine rinderpest has been described and depicted by MAURER et al. (1956), who were among the first modern observers to investigate the histopathology of the disease in cattle. They confirmed what was already well-established for the lesions caused by lapinised rinderpest virus in rabbits (FUKUSHO and NAKAMURA, 1940), i.e. that the virus has a particular predilection for the lymphoid tissues, in which location there is extensive destruction of lymphocytes in germinal centres, often accompanied by an increase of reticulo-endothelial elements, such as the macrophages. These observations were confirmed and amplified by independent studies published shortly afterwards (THIERY, 1956; KHERA, 1958a, b, c).

Multinucleated giant cells appeared in the lymphoid tissues on or about the 8th day after infection (KHERA, 1958a) and these contained eosinophilic, cytoplasmic inclusion bodies (Figs. 9 and 10) as well as degenerative nuclear changes which were regarded as specific (THIERY, 1956). PLOWRIGHT and FERRIS (1959a) found syncytia but no distinct inclusions in infected cattle, whilst MORNET et al. (1956b) found inclusions to be frequent in the lymphoid tissues of small ruminants. PLOWRIGHT (1962a) later found cytoplasmic inclusions to be common but intranuclear bodies were rare.

The stratified squamous epithelia, particularly those of the upper parts of the alimentary tract also showed syncytium formation which began in the *stratum spinosum* as foci of ballooning cells with acidophilic cytoplasmic inclusions (Figs. 7 and 8); there was also increased proliferative activity in the *stratum Malpighi*. Intranuclear inclusions were found according to KHERA (1958) or specific degenerative changes according to THIERY (1956). Similar lesions were seen in the mucosae of the vulva and prepuce and in the epidermis of the skin of the lips.

The changes described above are followed invariably by necrosis of the epithelium which detaches to leave a superficial erosion. Deeper ulcers, involving the submucosal or dermal tissues, do occur but are probably caused by secondary microbial invasion.

The mode of origin of the syncytial structures in infected lymphoid or epithelial tissues is not clear but THIERY (1956) found mitotic figures in those

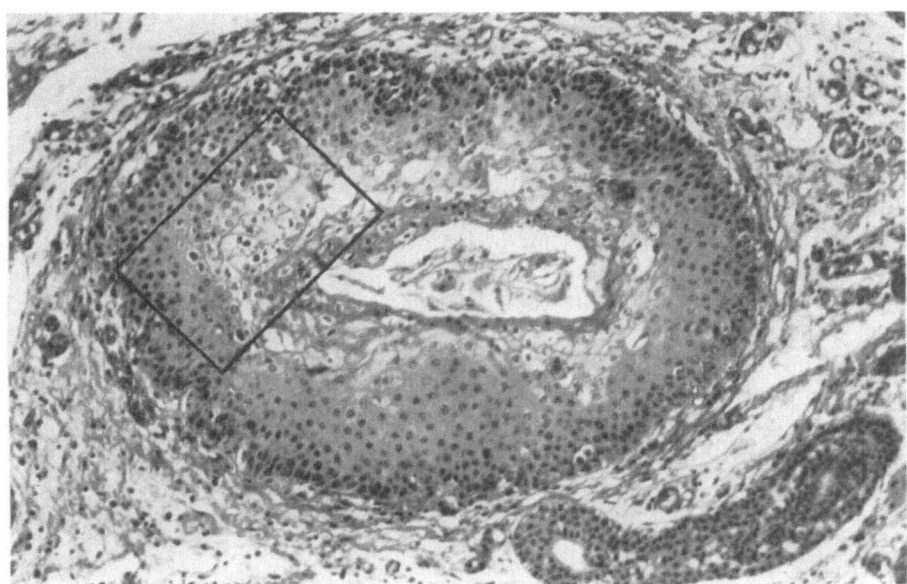

Fig. 7. Tonsillar crypt from an ox, killed on the 8th day after infection with low-passage culture virus of the RBOK strain. Note ballooning degeneration of the cells of the *stratum spinosum*. Zenker-formol: H and E (× 180).

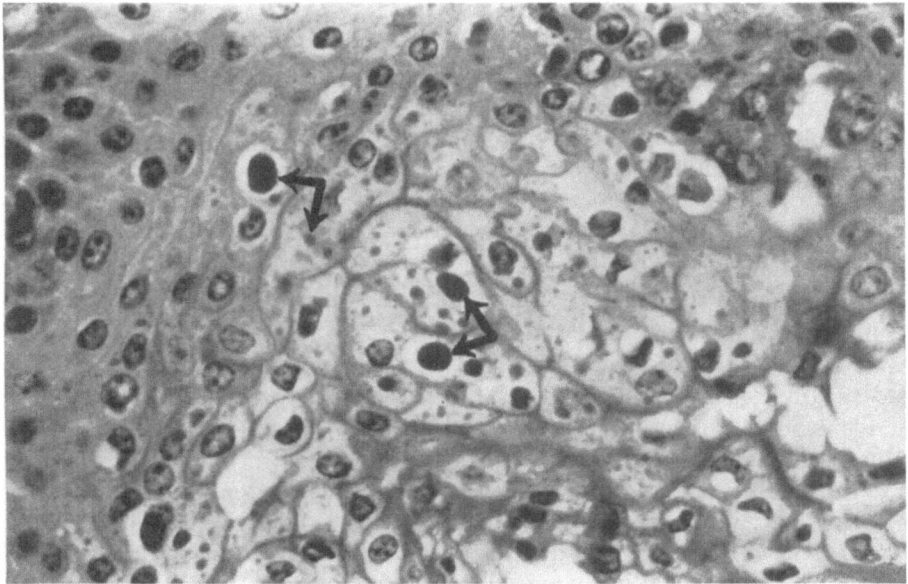

Fig. 8. Insert from Fig. 7 showing large and small intracytoplasmic inclusions (arrowed) (× 750).

of the mouth epithelia, implying nuclear without cytoplasmic division. KHERA (1958a), on the other hand, suggested that the syncytia with a peripheral ring of nuclei arose by the fusion of reticulo-endothelial elements.

3. Diagnosis of Rinderpest

In countries where the disease is enzootic rinderpest is usually diagnosed on clinico-pathological grounds, which are probably reliable enough if considered together with the herd history and the known presence or absence of diseases which simulate rinderpest in some of their clinical aspects, e.g. virus diarrhoea/ mucosal disease, papular stomatitis, infectious bovine rhinotracheitis, foot-and-mouth disease and malignant catarrhal fever. When rinderpest breaks out in areas which have hitherto been free of the disease, the inexperience of clinicians and the epizootiological importance of the occurrence virtually demand some accessory proof of the identity of the causal agent. Suitable techniques have been discussed in detail by MORNET and GILBERT (1960), SCOTT and BROWN (1961) and PROVOST and BORREDON (1963).

The laboratory confirmation of a diagnosis of rinderpest depends on one or more of the following, viz:

(a) The isolation of the virus from sick or dead animals;

(b) The detection of virus-specific antigens in their tissues;

(c) The demonstration of the development of antibodies;

(d) The histological examination of tissues for virus-specific changes.

These methods will each receive further brief mention:

a) In the past recovery of virus was carried out by the inoculation of blood or lymphoid tissues into susceptible and immune animals, particularly cattle or goats. To-day it is easier and cheaper to inoculate cultures of BK cells which have a sensitivity to field strains of virus which is comparable to that of cattle (PLOWRIGHT and TAYLOR — unpublished). In addition to its characteristic cytopathology, the virus can be identified by inhibition of its cytopathic effects with immune serum in a proportion of the cultures or by its later neutralisation (PLOWRIGHT, 1962b). Virus recovery by the inoculation of animals or tissue cultures always requires a longer time than could reasonably be allowed for the confirmation of outbreaks in "clean" areas.

b) The detection of virus-specific antigens, whether complement-fixing or precipitinogens reacting in agar-gel diffusion tests, allows a much more rapid answer to be obtained — in less than 24 hours in favourable circumstances. The agar-gel diffusion test is less sensitive than complement-fixation and has the disadvantage that the antigens are heat sensitive (SCOTT, 1962b). Both tests are ineffective very early in the course of the disease and also later after the appearance of antibody; nevertheless, infectious virus can be demonstrated at these times. It is also probable that the quantity of antigen in the tissues of animals depends directly on the virulence of the virus, which may be very low in enzootic areas (SCOTT and BROWN, 1961; PROVOST and BORREDON, 1963). Material for use in both tests can be obtained by lymph node biopsy (BROWN and SCOTT, 1960).

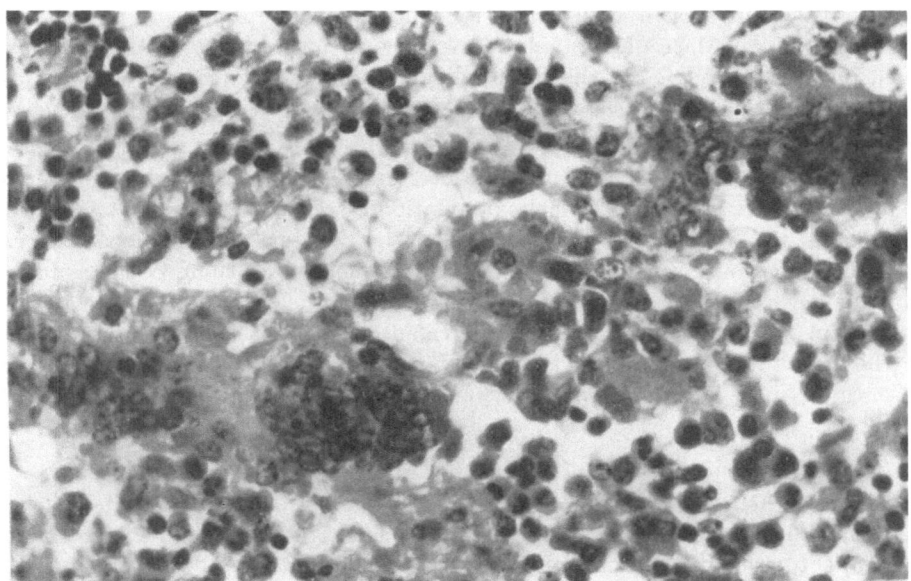

Fig. 9. Same animal as for Fig. 7. Haemolymph node showing large multinucleated formations. Zenker-formol: H and E (×735).

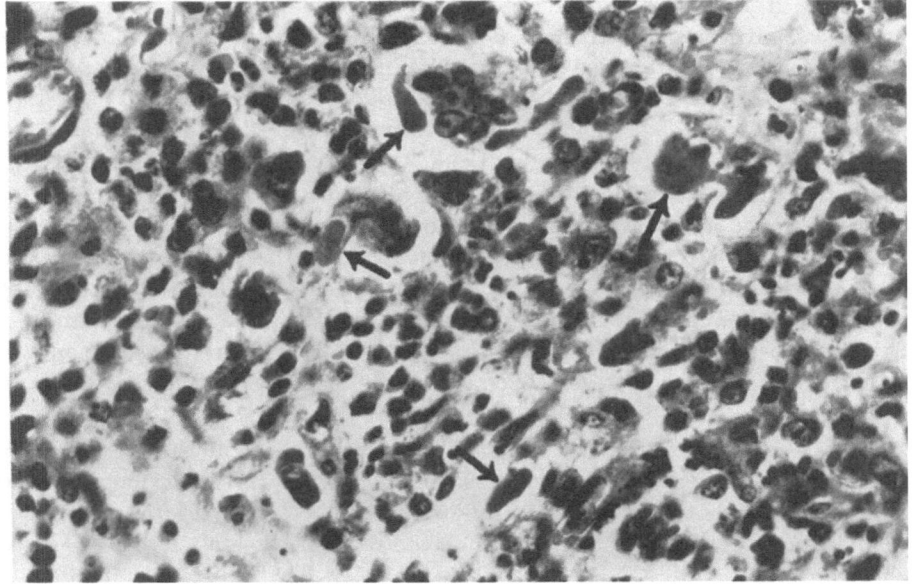

Fig. 10. Same animal as for Fig. 7. Lymphoid tissue of Peyer's patch with numerous, large, intracytoplasmic inclusions (arrowed). Zenker-formol: H and E (×680).

c) Demonstration of the development of antibodies requires acute-phase and convalescent sera in enzootic areas, whilst elsewhere it is perhaps unlikely that animals would be able or allowed to survive for sufficiently long periods. Neutralising antibodies are the most reliable although complement-fixing antibodies have been used successfully for diagnosis on a herd basis (NAKAMURA and MACLEOD, 1959; NAKAMURA, 1962).

d) It has been suggested (MORNET and GILBERT, 1960; PROVOST and BOR-REDON, 1963) that the histological examination of certain tissues, especially the tonsil and mucosae of the pharynx, oral cavity, vulva, prepuce, conjunctiva and nares, may offer an alternative or auxiliary method for rinderpest diagnosis. For this purpose it is necessary to demonstrate syncytia with cytoplasmic and/or nuclear inclusions. There is no published evidence for the efficacy of this approach to rinderpest diagnosis and it is unlikely to be accepted as conclusive in the absence of more specific confirmation, e.g. by fluorescent antibody (SCOTT, 1964).

K. Epizootiology

Perhaps the most important feature of the epizootiology of rinderpest is that virtually all recorded outbreaks in areas hitherto free of the disease have been due to the introduction of live, sick animals (RECEVEUR, 1952; SCOTT, 1957; ANON, 1966a) and not to indirect contagion as by infected meat, animal foodstuffs and feeding utensils or transporting vehicles. This is probably due in large measure to the fragility of the virus; stalls heavily contaminated by sick animals remained infective for a maximum of 48 hours in India according to SHILSTON (1917), for 4 days in Cairo (TODD and WHITE, 1914) and 3 but not 6 days in Kenya (MAURER, 1963). Heavily contaminated enclosures did not remain infective for more than 24 hours if devoid of vegetation and urine and faeces on grass only retained infectivity for 24—36 hours (WARD, WOOD and BOYNTON, 1914). In India, SHILSTON (1917) found that the infectivity persisted for only 6—8 hours in unshaded paddocks and for 18—24 hours if shaded.

Reacting cattle did not excrete enough virus into their environment to infect others during the first 48 hours of pyrexia (TODD and WHITE, 1914). Contact infection did not occur during the first 24 hours of fever or during convalescence — i.e. 10—20 days after experimental infection or more than 24 hours after the disappearance of pyrexia (WARD, WOOD and BOYNTON, 1914; COOPER, 1932). Infection by expired air could not be produced by cattle infected less than 5—6 days previously, i.e. before the 3rd to 4th days of pyrexia; transmission was thereafter successful up to death on the 10th day post-inoculation (IDNANI, 1944).

There is no evidence that vectors play any significant role in the dissemination of rinderpest although virus may persist in them for surprisingly long periods; among the possible transmitters the following have been investigated, viz: ticks, tabanid and stable flies, house flies, mosquitoes and sucking lice (CURASSON, 1921, 1932; BHATIA, 1935; MAURER, 1963); tsetse flies (HORNBY, 1926b) and leeches (NICOLLE and ADIL BEY, 1901; BOYNTON, 1913). In the latter virus remained viable for at least 25 days but transmission to cattle, by feeding infected leeches on them, was unsuccessful even after 15 minutes (BOYNTON, 1913).

Until recently, it was generally conceded that the dangers of importation of rinderpest with infected meat, even if frozen, were slight (SCOTT, 1957; RECEVEUR, 1957). The only reports which may be contrary to this opinion are those of ADEMOLLO (1958) concerning an incident in Italy in 1918 and a statement concerning fresh meat in Belgium in 1920 (ANON, 1920). The position has been changed by the recent demonstration that European pigs can acquire rinderpest by the *ingestion* of infected meat and that the virus can spread by contact from pig to pig and pig to cattle or *vice-versa* (SCOTT et al., 1959, 1962; DELAY and BARBER, 1962). It was particularly dangerous that the infection in pigs was sometimes so mild as to be easily overlooked (NICOLAS and RINJARD, 1921; SCOTT et al., 1959, 1962; DELAY et al., 1961) even after numerous passages in pigs (BARBER and HEUSCHELE, 1964). According to DELAY et al. (1961) the virus persisted in experimentally-infected pigs for as long as 36 days but this observation is so surprising as to warrant re-investigation.

Since rinderpest virus is so unstable outside the animal body and persists there for such a short time following the development of neutralising antibody, its maintenance in enzootic areas must depend on the continuing availability of fresh susceptible animals. The majority of adults in such areas are permanently immune, due to past infection, whether acquired naturally or from vaccination; calves are protected by maternal antibody for a period which varies according to the antibody titre of the dam's serum, but may last for 9—10 months. Hence, rinderpest tends to smoulder along in yearling animals only and in these the disease is often of a very mild type, especially in resistant breeds. This pattern has been established for many years; in the Masailands of Tanganyika it has persisted since at least the late 1920s (REID, 1949) and in central Africa since 1935 (PROVOST and BORREDON, 1963c).

These persistent enzootic foci are also often characterised in Africa by large populations of susceptible wild animals which help to maintain the virus and, by their uncontrolled movements, contribute to the wide dispersion of the disease (LOWE, 1942; THOMAS and REID, 1944; RECEVEUR, 1954). Control is made even more difficult in some years by the occurrence of widespread infection in wild ruminants without any detectable morbidity or mortality (PLOWRIGHT, 1963c; PLOWRIGHT and McCULLOCH, 1967). In south-east Asia domestic pigs and wild swine and ruminants may play a similar role (see SCOTT, 1964) and in other areas there is increasing evidence for mild or subclinical infection, by rinderpest or a closely-related virus, in small domestic ruminants (ZWART and ROWE, 1966).

It has been clearly shown in recent experiments that Nigerian field isolates can spread, although irregularly, by close contact from cattle to sheep and goats and then amongst the small ruminants (ZWART and MACADAM, 1967a) but extension from sheep to cattle was not demonstrable and from goats to cattle it was infrequent (ZWART and MACADAM, 1967b). It was concluded that the immunisation of sheep and goats was probably unnecessary once rinderpest had been eliminated from cattle (ZWART and ROWE, 1967b); this is a matter of considerable importance, since in India alone it was estimated that the need to immunise sheep and goats would increase the number of vaccinations required from about 140 millions over 5 to 10 years (ANON, 1965) to 320 millions (DATTA, 1954).

In countries which have been free of the disease for long periods or which have never been infected, rinderpest can be rapidly and completely eliminated by quarantine and vaccination, as in the Philippines in 1955 (CORONEL, 1960), by movement restrictions and slaughter as in Brazil, 1920 and Australia, 1923 (ROBERTS, 1921; ROBERTSON, 1924), or quarantine, slaughter and antiserum as in Belgium, 1920 (ANON, 1920).

References

ADAMS, J. M.: Comparative study of canine distemper and a respiratory disease of man. Pediatrics 11, 15—25 (1953).

ADAMS, J. M., and D. T. IMAGAWA: Immunological relationships between measles and distemper viruses. Proc. Soc. exp. Biol., (N.Y.) 96, 240—244 (1957).

ADEMOLLO, A.: A proposito di importazione di carne dai paesi infetti di peste bovina. Vet. ital. 9, 130—138 (1958).

ANDERSON, C. D., and J. G. ATHERTON: Effect of actinomycin D on measles virus growth and interferon production. Nature (Lond.) 203, 670—671 (1964).

ANDREWES, C. H. (Chairman): Proposals and Recommendations of the Provisional Committee for Nomenclature of Viruses (1965).

ANON.: La peste povine en Belgique. Rev. gén. Méd. vét. 29, 577—583 (1920).

ANON.: Report by Delegation of India: Rinderpest (Prophylaxis and Eradication Campaign) in India. Bull. Off. int. Epiz. 63, 47—55 (1965).

ANON.: Connaissances acquises récemment sur la peste bovine et son virus. Rev. Elev. 19, 365—413 (1966a).

ANON.: Laboratoire de Farcha. Rapport Annuel, 1965 (1966b).

BAKER, J. A.: Rinderpest VIII. Rinderpest infection in rabbits. Amer. J. vet. Res. 7, 179—182 (1946).

BAKER, J. A., and A. S. GREIG: Rinderpest XII. The successful use of young chicks to measure the concentration of rinderpest virus propagated in eggs. Amer. J. vet. Res. 7, 196—198 (1946).

BAKER, J. A., J. TERRENCE, and A. S. GREIG: Rinderpest X. Response of guinea pigs to the virus of rinderpest. Amer. J. vet. Res. 7, 189—192 (1946).

BARBER, T. L., and W. P. HEUSCHELE: Experimental passage of rinderpest virus in North American pigs. Bull. epizoot. Dis. Afr. 12, 277—285 (1964).

BAWA, H. S.: Rinderpest in sheep and goats in Ajmer Merwara. Ind. J. vet. Sci. 10, 103—112 (1940).

BEATON, W. G.: Rinderpest in goats in Nigeria. J. comp. Path. 43, 301—307 (1930).

BHATIA, H. L.: The role of Tabanus Orientis (Wlk) and Stomoxys calcitrans (Linn.) in the mechanical transmission of rinderpest. Ind. J. vet. Sci. 5, 2—22 (1935).

BIRKETT, J. D.: Duration of immunity conferred by wet lapinised rinderpest virus vaccine in Ndama cattle in Sierra Leone. J. comp. Path. 68, 115—120 (1958).

BLACK, F. L.: Relationship between virus particle size and filterability through Gradacol membranes. Virology 5, 391—392 (1958).

BLACK, F. L., M. REISSIG, and J. MELNICK: Measles Virus. Advanc. Virus Res. 6, 205—227 (1959).

BÖGEL, K., G. ENDERS-RUCKLE et A. PROVOST: Une réaction sérologique rapide de mesure des anticorps antibovipestiques. C. R. Acad. Sci. (Paris) 259, 482—484 (1964).

BOLDRINI, G.: Un episodio di peste bovina su una nave del „Lloyd Triestino". Vet. ital. 5, 1182—1183 (1954).

BOULANGER, P.: The use of the complement-fixation test for the demonstration of rinderpest virus in rabbit tissue using rabbit antisera. Canad. J. comp. Med. 21, 363—369 (1957a).

BOULANGER, P.: Application of the complement-fixation test to the demonstration of rinderpest virus in the tissue of infected cattle using rabbit antiserum: I. Results with the Kabete and Pendik strains of virus. Canad. J. comp. Med. 21, 379—388 (1957b).

BOYNTON, W. H.: Duration of the infectiveness of virulent rinderpest blood in the water leech, *Hirudo Boyntoni*, Wharton. Philipp. J. Sci. (B) 509—521 (1913).

BOYNTON, W. H.: A preliminary report of experiments on the cultivation of the virus of rinderpest *in vitro*. Philipp. J. Sci. (B) 9, 39—44 (1914).

BOYNTON, W. H.: Rinderpest in swine with experiments upon its transmission from cattle and carabaos to swine and *vice versa*. Philipp. J. Sci. 11, 215—263 (1916).

BREESE, S. S., and C. J. DE BOER: Electron microscopy of rinderpest virus in bovine kidney tissue culture cells. Virology 19, 340—348 (1963).

BRENNER, S., and R. W. HORNE: A negative staining method for high resolution electron microscopy of viruses. Biochim. biophys. Acta (Amst.) 34, 103—110 (1959).

BRION, G., et J. GRUEST: Isolement et maintien *in vitro* d'une souche essentielle-ment constituée de cellules épithéliales et obtenue a partir d'un rein de bovidé. Ann. Inst. Pasteur 92, 426—429 (1957).

BRISTOWE, J. S.: On the morbid anatomy of the cattle plague. In 3rd Report of the Commissioners appointed to inquire into the origin and the nature, etc. of the cattle plague. pp. 81—127. London: Her Majesty's Stationary Office (1866).

BROTHERSTON, J. G.: Lapinised rinderpest virus and a vaccine; some observations in East Africa. 1. Laboratory experiments. J. comp. Path. 61, 263—288 (1951a).

BROTHERSTON, J. G.: Lapinised rinderpest virus and a vaccine; some observations in East Africa. II. Field trials with lapinised vaccine. J. comp. Path. 61, 289—306 (1951b).

BROTHERSTON, J. G.: Rinderpest: some notes on control by modified virus vaccines. I. Vet. Rev. Annot. 2, 95—106 (1956).

BROTHERSTON, J. G.: Rinderpest: some notes on control by modified virus vaccines. II. Vet. Rev. Annot. 3, 45—56 (1957).

BROTHERSTON, J. G.: Rinderpest: some notes on control by modified virus vaccines. III. Vet. Rev. Annot. 4, 49—54 (1958).

BROUDIN, L. (1923): Quoted by CURASSON (1932; 1942).

BROWN, R. D.: Rinderpest immunity in calves — a review. Bull. epizoot. Dis. Afr. 6, 127—133 (1958a).

BROWN, R. D.: Rinderpest immunity in calves. I. The acquisition and persistence of maternally-derived antibody. J. Hyg. (Lond.) 56, 427—434 (1958b).

BROWN, R. D.: Rinderpest immunity in calves. II. Active immunisation. J. Hyg. (Lond.) 56, 435—444 (1958c).

BROWN, R. D., and A. RASCHID: Duration of rinderpest immunity in cattle following vaccination with caprinised rinderpest virus. Bull. epizoot. Dis. Afr. 13, 311—315 (1965).

BROWN, R. D., and G. R. SCOTT: A screening procedure for the detection of rinderpest immune cattle. Bull. epizoot. Dis. Afr. 7, 169—171 (1959).

BROWN, R. D., and G. R. SCOTT: Diagnosis of rinderpest by lymph node biopsy. Vet. Rec. 72, 1055—1056 (1960).

BUSSELL, R. H., and D. T. KARZON: Canine distemper virus in chick embryo cell cul-ture. Plaque assay, growth and stability. Virology 18, 589—600 (1962).

CACCAVELLA, A. E.: Observations sur la transmission de la peste bovine chez les anti-lopes pongo *(Tragelaphus scriptus)* et isha *(Silvicapra grimmi)*. Sérothérapie et virus de passage sur les bovidés. Ann. Soc. belge. Méd. trop. 16, 309—312 (1936).

CARLSTROM, G.: Neutralisation of canine distemper virus by serum of patients con-valescent from measles. Lancet 2, 344 (1957).

CARMICHAEL, J.: The virus of rinderpest in relation to *Glossina morsitans* (Weston). Bull. ent. Res. 24, 337—342 (1933).

CARMICHAEL, J.: Rinderpest in African game. J. comp. Path. 51, 264—268 (1938a).

CARMICHAEL, J.: In A. R. Vet. Dept., Uganda, 1937. pp. 19—20 (1938b).

CARMICHAEL, J.: In A. R. Vet. Dept., Uganda, 1938. pp. 21—23 (1939).

CARRÉ et FRAIMBAULT: Note sur la contagiosité de la peste bovine au porc. Ann. Inst. Pasteur 12, 848—856 (1898).

CARTER, G. R. (1952): Quoted by McKERCHER, 1957.

CARTER, G. R., and C. A. MITCHELL: Method for adapting the virus of rinderpest to rabbits. Science **128**, 252—253 (1958).

CASALS, J., P. OLITSKY, and R. O. ONSLOW: A specific complement-fixation test for infection with poliomyelitis virus. J. exp. Med. **94**, 123—137 (1951).

CEBE, J., and J. PERRIN: Conservation du virus de la peste bovine par passages sur le lapin. Bull. econ. Indochin. **38**, 795—798 (1935). Abstr. in Vet. Bull. **6**, 744 (1936).

CHENG, S. C., T. C. CHOW, and H. R. FISCHMAN: Avianised rinderpest vaccine in China. In Rinderpest vaccines, their production and use in the field. FAO Agricultural Studies No. 8, pp. 29—39 (1949).

CHENG, S. C., and H. R. FISCHMAN: Lapinised rinderpest virus. In Rinderpest vaccines, their production and use in the field. FAO Agricultural Studies No. 8, pp. 47—63 (1949).

CILLI, V.: Atteggiamenti biologici del virus KAG (Kabete attenuated goat) della peste bovina sulle capre Eritree. Revista di Biologia **43**, 255—290 (1951).

CILLI, V., V. MAZZARACCHIO e C. ROETTI: L'epidosio di peste bovina al Giardino Zoologico di Roma. Arch. ital. Sci. med. trop. **32**, 83—64 (1951). Quoted by SCOTT (1964).

COOPER, H.: Rinderpest: transmission of infection by contact. Ind. J. vet. Sci. **2**, 384—360 (1932).

COOPER, H. K.: Rinderpest XVI. Complement-fixation test for rinderpest. Amer. J. vet. Res. **7**, 228—237 (1946).

COOPER, P. D.: A chemical basis for the classification of animal viruses. Nature (Lond.) **190**, 302—305 (1961).

CORNELL, R. L.: In A. R. Vet. Dept., Tanganyika, 1933. pp. 37—42 (1934).

CORNELL, R., and R. OONYWONGSE: Observations on rinderpest immunisation with goat virus. Ind. J. vet. Sci. **11**, 1—15 (1941).

CORONEL, A. B.: Rinderpest in the Philippines. Bull. Off. int. Epiz. **53**, 100—103 (1960).

COWDRY, E. V.: The problem of intranuclear inclusions in virus diseases. Arch. Path. **18**, 527—542 (1934).

CRAWFORD, M.: The immunology and epidemiology of some virus diseases. Vet. Rec. **59**, 537—540 (1947).

CRUIKSHANK, J. G., A. P. WATERSON, A. D. KANAREK, and D. M. BERRY: The structure of canine distemper virus. Res. Vet. Sci. **3**, 485—486 (1962).

CURASSON, G.: Notes sur la peste bovine en Afrique Occidentale Française et en Pologne. Rev. gén. Méd. vét. **30**, 569—590 (1921).

CURASSON, G.: Les séquelles de la peste bovine et les porteurs de germes. Rev. gén. Méd. vét. **34**, 549—554 (1926).

CURASSON, G.: La Peste Bovine. Vigot Frères, Paris (1932).

CURASSON, G.: Traité de Pathologie Exotique Vétérinaire et Comparée. Vol. I. pp. 12—169. Vigot Frères, Paris (1942).

DATTA. S.: The national rinderpest eradication plan. Ind. J. vet. Sci. **24**, 1—9 (1954).

DATTA, S. C. A., and V. R. RAJAGOPALAN: An unusual case of chronic rinderpest with special reference to the carrier problem in this disease. Ind. J. vet. Sci. **2**, 357—382 (1932).

DAUBNEY, R.: Observations on rinderpest. J. comp. Path. **41**, 228—248 (1928).

DAUBNEY, R.: In A. R. Agric. Dept., Kenya, 1936, **2**, 63—66 (1937a).

DAUBNEY, R.: Rinderpest: a résumé of recent progress in East Africa. J. comp. Path. **50**, 405—409 (1937b).

DAUBNEY, R.: Newer researches regarding tropical and subtropical diseases. Proc. XIIIth Int. vet. Congr., Zurich-Interlaken, 21—33 (1938).

DAUBNEY, R.: In A. R. Vet. Dept., Kenya. 1642, p. 5 (1943).

DAUBNEY, R.: Recent developments in rinderpest control. Bull. Off. int. Epiz. **28**, 36—45 (1947).

DAUBNEY, R.: Immunisation against rinderpest by means of the goat-adapted virus. Proc. 4th Int. Congr., trop. Med. Malaria. Washington **2**, 1358—1365 (1948).

DAUBNEY, R.: In "Rinderpest Vaccines; their production and use in the field". Ed. K. V. L. KESTEVEN. pp. 6—18. Food and Agriculture Organization of the United Nations (1949).

DAUBNEY, R.: Peste Bovine. Notes sur les vaccinations par virus vivants. Bull. Off. int. Epiz. **36**, 116—128 (1951).

D'COSTA, J., and B. SINGH: Rinderpest: clinical syndrome in goats in India. Ind. J. vet. Sci. **3**, 122—128 (1933).

DEBOER, C. J.: Adaptation of two strains of rinderpest virus to tissue culture. Fed. Proc. **19**, 405 (1960).

DEBOER, C. J.: Adaptation of two strains of rinderpest virus to tissue culture. Arch. ges. Virusforsch. **11**, 534—543 (1961).

DEBOER, C. J., and T. L. BARBER: pH and thermal stability of rinderpest virus. Arch. ges. Virusforsch. **15**, 98—108 (1964).

DELAY, P. D., and T. L. BARBER: Transmission of rinderpest virus from experimentally-infected cattle to pigs. Proc. U.S. Livestock Sanitary Assoc., 66th Ann. Mtg. pp. 132—136 (1962).

DELAY, P. D., W. M. MOULTON, and S. S. STONE: Survival of rinderpest virus in experimentally-infected swine. Proc. U.S. Livestock Sanitary Assoc., 65th Ann. Mtg. pp. 376—383 (1961).

DELAY, P. D., S. S. STONE, D. T. KARZON, S. KATZ, and J. ENDERS: Clinical and immune response of alien hosts to inoculation with measles, rinderpest and canine distemper viruses. Amer. J. vet. Res. **26**, 1359—1373 (1965).

DELPY, L.: Peste bovine atypique transmise par passage. Contribution a l'étude des porteurs de virus. Rev. gén. Méd. vét. **37**, 259—264 (1928).

DELPY, L.: La peste bovine en Perse. Rev. gén. Méd. vét. **44**, 577—601 (1935).

DHANDA, M. R., and S. L. MANJREKAR: Observations on rinderpest amongst sheep and goats in the State of Bombay. Ind. vet. J. **28**, 306—319 (1952).

DHILLON, S. S.: Incidence of rinderpest in camels in Hissar district. Ind. vet. J. **36**, 603—607 (1959).

DIECKERHOFF, W.: Geschichte der Rinderpest und ihrer Literatur. Beitrag zur Geschichte der vergleichenden Pathologie. Enslin, Berlin (1890).

DREW, R. M.: Isolation and propagation of rabbit epithelial-like cells. Science **126**, 747—748 (1957).

EDWARDS, J. T.: Rinderpest: some properties of the virus and further indications for its employment in the serum-simultaneous method of protective inoculation. Transactions of the Far Eastern Association of Tropical Medicine. 7th Congress held in India. **3**, 699—706 (1927a).

EDWARDS, J. T.: Rinderpest: some points in immunity. Transactions of the Far Eastern Association of Tropical Medicine. 7th Congress held in India. **3**, 707—717 (1927b).

EDWARDS, J. T.: The problem of rinderpest in India. Bull. No. 199, Imperial Institute of Agricultural Research, Pusa. Govt. of India, Central Publication Branch, Calcutta (1930).

ENDERS, J. F.: Cytopathology of virus infections. Ann. Rev. Microbiol. **8**, 473—502 (1954).

FRANKLIN, R. M., and E. WECKER: Inactivation of some animal viruses by hydroxylamine and the structure of ribonucleic acid. Nature (Lond.) **184**, 343—345 (1959).

FRY, R. M., and R. I. N. GREAVES: The survival of bacteria during and after drying. J. Hyg. (Lond.) **49**, 220—246 (1951).

FUKUSHO, K., and K. FURUYA: Studies on simultaneous inoculation of anti-rinderpest serum and lapinised rinderpest virus into the Japanese Black cattle; in Exp. Rep. No. 26, Govt. exp. Stat. anim. Hyg., Tokyo. pp. 27—34 (1953).

FUKUSHO, K., and J. NAKAMURA: On the experimental infection with rinderpest virus in the rabbit. II. Pathology. Jap. J. vet. Sci. **2**, 75—101 (1940). Abstr. in Vet. Bull. **11**, 163 (1941).

FURUTANI, T., S. ISHII, K. KURATA, and H. NAKAMURA: Studies on the AKO strain of lapinised-avianised rinderpest virus. II. Features of multiplication of the virus in embryonated hens' eggs. Bull. nat. Inst. anim. Hlth., Tokyo **32,** 137—149 (1957 b).

FURUTANI, T., T. KATAOKA, K. KURATA, and H. NAKAMURA: Studies on the AKO strain of lapinised-avianised rinderpest virus. I. Avianisation of lapinised rinderpest virus. Bull. nat. Inst. anim. Hlth., Tokyo **32,** 117—135 (1957 a).

FURUTANI, T., H. NAKAMURA, S. ISHII, and K. KURATA: Studies on the rinderpest virus neutralisation method in embryonating eggs. I. Route of inoculation. Jap. J. vet. Sci. **16,** 56—57 (1954). Abstr. in Vet. Bull. **26,** 380 (1956).

GAMGEE, J.: "The Cattle Plague". Hardwicke, London (1866).

GARGADENNEC, L., et A. LALANNE: La peste des petits ruminants. Bull. Serv. zootech. Epiz. A.O.F. **8,** 16—21 (1942). Quoted by ANON (1966 a).

GIBBS, C. S.: Filterable virus carriers. J. infect. Dis. **53,** 169—174 (1933).

GILBERT, Y., et J. MONNIER: Adaptation d'une souche de virus bovipestique à la culture cellulaire. Rev. Élev. **15,** 311—320 (1962 a).

GILBERT, Y., et J. MONNIER: Adaptation du virus de la peste des petits ruminants aux cultures cellulaires. Rev. Élev. **15,** 321—335 (1962 b).

GILBERT, Y., P. MORNET et Y. GOUEFFON: Comportement humoral du bœuf et du lapin envers l'inoculation de virus de Carré: ses rapports avec l'immunisation contre le virus bovipestique normal ou modifiée. Bull. Acad. vét. Fr. **33,** 305—315 (1960).

GIRARD, H., et M. CHARITAT: La vaccination antipestique au Soudan a l'aide du virus pestique caprin. Rév. Élev. **1,** 7—15 (1947).

GORET, P., A. BRION et M. FONTAINE: Echec des essais de prévention et de traitement de la maladie de Carré par le sérum contre la peste bovine. Bull. Acad. vét. Fr. **33,** 343—347 (1960 c).

GORET, P., J. FONTAINE, C. MACKOWIAK, C. PILET et T. CAMARA: Neutralisation du virus de la maladie de Carré par le sérum contre la peste bovine. Bull. Acad. vét. Fr. **32,** 287—296 (1959).

GORET, P., P. MORNET, Y. GILBERT et C. PILET: Immunité croisée entre la maladie de Carré et la peste bovine. Bull. Acad. vét. Fr. **31,** 163—166 (1958) or C.R. Acad. Sci. (Paris) **245,** 2564—2566 (1957).

GORET, P., P. MORNET, Y. GILBERT, C. PILET et G. ORTH: Recherches sur l'immunisation croisée "maladie de Carré — peste bovine" chez le lapin. Ann. Inst. Pasteur **98,** 605—610 (1960 a).

GORET, P., C. PILET, M. GIRARD et T. CAMARA: Apparition et durée de l'immunité contre la maladie de Carré conferée au furet par le virus lapinisé de la peste bovine. Ann. Inst. Pasteur **98,** 610—612 (1960 b).

GRIEFF, D., W. A. RIGHTSEL, and E. E. SCHULER: Effect of freezing, storage at low temperatures and drying by sublimation in vacuo on the activities of measles virus. Nature (Lond.) **202,** 624—625 (1964).

GUPTA, K. C. S., and N. S. VERMA: Rinderpest in wild ruminants. Ind. J. vet. Sci. **19,** 219—222 (1949).

GUYAUX, R.: Gibier et peste bovine; cas de transmission de la peste bovine du buffle au bétail bovin. Bull. agric. Congo Belge **42,** 123—129 (1951).

HADDOW, J. R., and J. A. IDNANI: The vaccination of highly susceptible animals against rinderpest. Ind. J. vet. Sci. **17,** 1—10 (1947).

HALE, M. W., and R. V. L. WALKER: Rinderpest. XIII. The production of rinderpest vaccine from an attenuated strain of virus. Amer. J. vet. Res. **7,** 199—211 (1946).

HALE, M. W., R. V. L. WALKER, F. D. MAURER, J. A. BAKER, and D. L. JENKINS: Rinderpest. XIV. Immunisation experiments with attenuated rinderpest vaccine including some observations on the keeping qualities and potency tests. Amer. J. vet. Res. **7,** 212—221 (1946).

HALL, G. N.: Studies on rinderpest immunisation. Thesis, Zurich (1933).

HAYFLICK, L., and P. S. MOORHEAD: The serial cultivation of human diploid cell strains. Exp. Cell Res. **25**, 585—621 (1961).

HENDERSON, W. M., and J. B. BROOKSBY: The survival of foot-and-mouth disease virus in meat and offal. J. Hyg. (Lond.) **46**, 394—402 (1948).

HENDERSON, W. W.: Rinderpest immunisation by means of goat attenuated virus. A. R. Vet. Dept., Nigeria, 1943, pp. 21—25 (1945).

HENLE, G., and F. DEINHARDT: Propagation and primary isolation of mumps virus in tissue culture. Proc. Soc. exp. Biol. (N.Y.) **89**, 556—560 (1955).

HENNING, M. W.: Rinderpest — in Animal Diseases in South Africa, 3rd Ed., Central News Agency, Ltd., South Africa (1956).

HILSONT, P., et C. BOURDEREAUX: Une enzootie pestique cryptogénétique sur des phacochères en captivité à Bamako (Soudan française). Rev. Élev. **7**, 79—80 (1954).

HOLMES, F. O.: In "Bergey's Manual of Determinative Bacteriology", 6th Ed., p. 1227; Williams and Wilkins, Baltimore, Maryland, U.S.A. (1948).

HORNBY, H. E.: In A. R. Vet. Dept., Tanganyika, 1925, p. 30 (1926a).

HORNBY, H. E.: Studies in rinderpest immunity. (2) Methods of infection. Vet. J. **82**, 348—355 (1926b).

HORNBY, H. E.: The distribution of rinderpest virus in infected blood. J. comp. Path. **41**, 17—24 (1928).

HUARD, M., J. ANDRE et J. FOURNIER: Essais de titrage des anticorps neutralisant le virus bovipestique. Ann. Inst. Pasteur **96**, 506—509 (1959).

HUDSON, J. R.: Rinderpest virus attenuated in eggs. Vet. Rec. **59**, 331 (1947).

HUDSON, J. R., and W. B. C. DANKS: In A. R. Vet. Dept., Kenya 1947, pp. 13—14 (1949).

HUSSAIN, S. F., and M. M. SARWAR: Value of gel diffusion precipitation reaction in detection of rinderpest vaccinated cattle. W. Pakist. J. agric. Res. **1**, 147—151 (1962). Abstr. in Vet. Bull. **34**, 401 (1964).

HUTCHEON, D.: Rinderpest in South Africa. J. comp. Path. **15**, 300—324 (1902).

HUYGELEN, C.: Runderpestvirus in Weefselkultur. Vlaams Diergeneesk. T. **29**, 46—53 (1960a).

HUYGELEN, C.: Failure to demonstrate agglutination of red cells by rinderpest virus. Bull. epizoot. Dis. Afr. **8**, 121—125 (1960b).

IDNANI, J. A.: Transmission of rinderpest by expired air. Ind. J. vet. Sci. **14**, 216—220 (1944).

ILLARTEIN, P. R., et M. GUERRET: Contribution à l'étude de la prophylaxie de la peste en Guinée-française (A.O.F.). Note sur les essais de vaccination de taurins N'dama au moyen de la souche de virus pestique lapinisé Nakamura III. Bull. Soc. Path. exot. **47**, 422—434 (1954).

IMAGAWA, D. T.: Propagation of rinderpest virus in suckling mice and its comparison to murine adapted strains of measles and distemper. Arch. ges. Virusforsch. **17**, 203—215 (1965).

IMAGAWA, D. T., P. GORET, and J. M. ADAMS: Immunological relationships of measles, distemper and rinderpest viruses. Proc. nat. Acad. Sci. (Wash.) **46**, 1119—1123 (1960).

INOUE, T.: On the rabbit passage of rinderpest virus. J. Jap. Soc. vet. Sci. **13**, 314—316 (1934).

INOUE, T., S. HARADA, and T. SHIMIZU: Preliminary note on the experimental infection with rinderpest virus in susliks. Selected Contributions from the Mukden Institute for Infectious Diseases of Animals **1**, 221—222 (1930). Abstr. in Vet. Bull. **1**, 53 (1931).

ISHII, S., G. TOKUDA, and M. WATANABE: Analysis of rinderpest virus antigen I. Results of the diffusion precipitation test in agar-gel. Nat. Inst. anim. Hlth. Quart. **4**, 205—213 (1964).

ISHII, S., and K. TSUKUDA: Studies on the adaptation of bovine strain rinderpest in chick embryos. Exp. Rep. Govt. exp. Stat. anim. Hyg. (Tokyo) **25**, 29—36 (1952).

ISOGAI, S.: On the rabbit virus inoculation as an active immunisation method against rinderpest for Mongolian cattle. Jap. J. vet. Sci. **6,** 388—390 (1944).

ISOGAI, S.: Pathogenicity of rinderpest virus, original and attenuated, in various tissue cultures. J. Jap. Ass. inf. Dis. **55,** 417—432 (1961). Quoted by TOKUDA et al. (1962).

IYER, S. V., and R. SRINAVASAN: Studies on a new Madras strain of lapinised rinderpest virus suitable for use in vaccine production. Ind. vet. J. **31,** 155—184 (1954).

JACKSON, R., et D. A. E. CABOT: La résistance du virus de la peste bovine. Bull. Off. int. Epiz. **3,** 775—783 (1930).

JACOBSON, W., and M. WEBB: The two types of nucleoprotein during mitosis. Exp. Cell Res. **3,** 163—183 (1952).

JACOTOT, H.: Sur l'état d'infection inapparente dans la peste bovine: conséquences épidémiologiques possible. Bull. Soc. Path. exot. **22,** 239—241 (1929).

JACOTOT, H.: Sur la sensibilité du lapin au virus de la peste bovine. Bull. Soc. Path. exot. **23,** 904—909 (1930).

JACOTOT, H.: Sur la teneur en virus de quelques tissues des veaux atteints de peste bovine expérimentale. Bull. Soc. Path. exot. **24,** 21—26 (1931a).

JACOTOT, H.: Existe-t-il en Indochine des porteurs de virus pestique. Bull. Soc. Path. exot. **24,** 51—58 (1931b).

JACOTOT, H.: L'infection pestique qui entraine l'avortement peut-elle être propagée par le foetus et la femelle qui l'a expulsé. Bull. Soc. Path. exot. **24,** 74—76 (1931c).

JACOTOT, H.: Existe-t-il en Indochine plusieurs virus pestiques. Bull. Soc. Path. exot. **24,** 521—526 (1931d).

JACOTOT, H.: Études sur la peste bovine. I. Recherches sur le virus de la peste bovine et sur l'infection qu'il détermine. Ann. Inst. Pasteur **48,** 377—399 (1932a).

JACOTOT, H.: Observations et recherches sur la peste bovine du bétail d'Indochine. Arch. Inst. Pasteur d'Indochine **15,** 7—95 (1932b).

JACOTOT, H.: Rapport concernant le contrôle et la standardisation des sérums et vaccins contre la peste bovine. Bull. Off. int. Epiz. **33,** 168—183 (1950).

JENKINS, D. L., and R. E. SHOPE: Rinderpest. VII. The attenuation of rinderpest virus for cattle by cultivation in embryonated eggs. Amer. J. vet. Res. **7,** 174—178 (1946).

JENKINS, D. L., and R. V. L. WALKER: Rinderpest IX. Neutralisation tests in rabbits as a measure of the immune responses in calves to vaccination against rinderpest. Amer. J. vet. Res. **7,** 183—188 (1946).

JOHNSON, R. H.: An outbreak of rinderpest involving cattle and sheep. Vet. Rec. **70,** 457—461 (1958).

JOHNSON, R. H.: Rinderpest in tissue culture. I. Methods for virus production. Brit. vet. J. **118,** 107—116 (1962a).

JOHNSON, R. H.: Rinderpest in tissue culture. II. Serum neutralisation tests. Brit. vet. J. **118,** 133—140 (1962b).

JOHNSON, R. H.: Rinderpest in tissue culture. III. Use of the attenuated strain as a vaccine for cattle. Brit. vet. J. **118,** 141—150 (1962c).

KAKIZAKI, C.: Study on the glycerinated rinderpest vaccine. Kitasato Arch. exp. Med. **2,** 59—66 (1918). Quoted by SCOTT (1964).

KELSER, R. A., S. YOUNGBERG, and T. TOPACIO: An improved vaccine for immunisation against rinderpest. J. Amer. vet. med. Assoc. **74,** 28 (1929).

KHERA, K. S.: Étude histologique de la peste bovine. I. Pathogénèse du virus de la peste bovine dans les ganglions lymphatiques. Rev. Élev. **11,** 399—405 (1958a).

KHERA, K. S.: Étude histologique de la peste bovine. II. Lésions histologiques au niveau du tracts digestif. Rev. Élev. **11,** 406—415 (1958b).

KHERA, K. S.: Étude histologique de la peste bovine. III. Lésions dans les différents organes. Rev. Élev. **11,** 416—420 (1958c).

KINLOCH, B. G.: In A. R. Game Division, Tanganyika, 1961 (1963).

KOCH, R.: Berichte des Herrn Prof. Dr. KOCH über seine in Kimberley gemachten Versuche bezüglich Bekämpfung der Rinderpest. Zbl. Bakt. I. Abt. Orig. **21,** 526—537 (1897).

KOLLE, W.: Beiträge zur Klärung der Frage über die Wirkungsweise der Rinderpest-galle. Z. Hyg. Infekt.-Kr. **30**, 33—46 (1899).

KOPROWSKI, H.: Counterparts of human viral disease in animals. Ann. N.Y. Acad. Sci. **70**, 369—381 (1958).

KUNERT, H.: Züchtung des Rinderpestvirus auf der Chorio-Allantois des Hühner-embryo. Dtsch. tierärztl. Wschr. **46**, 487—490 (1938).

KYLASAMAIER, K.: A study of Madras fowl pest. Ind. vet. J. **7**, 340—346 (1931).

LALL, H. K.: Some observations on the immunisation of sheep and goats against rinderpest. Ind. J. vet. Sci. **17**, 11—22 (1947).

LAM, K. S. K., and J. G. ATHERTON: Measles virus. Nature (Lond.) **197**, 820—821 (1963).

LECLAINCHE, E.: Histoire de la Médicine Vétérinaire. Office du Livre, Toulouse (1936).

LEVINE, S., and W. OLSON: Nucleic acids of measles and vesicular stomatitis viruses. Proc. Soc. exp. Biol. (N.Y.) **113**, 630—631 (1963).

LIBEAU, J., and G. R. SCOTT: Rinderpest in eastern Africa to-day. Bull. epizoot. Dis. Afr. **8**, 23—26 (1960).

LIESS, B.: Fluoreszenzserologische Untersuchungen an Zellkulturen nach Infektion mit Rinderpestvirus. Zbl. Bakt. I. Abt. Orig. **190**, 424—444 (1963).

LIESS, B.: Untersuchungen über das Virus der Rinderpest unter Verwendung von Zellkulturen. Arch. exp. Vet.-Med. **20**, 157—257 (1964).

LIESS, B., and W. PLOWRIGHT: Studies in tissue culture on the pH stability of rinder-pest virus. J. Hyg. (Lond.) **61**, 205—211 (1963a).

LIESS, B., and W. PLOWRIGHT: The propagation and growth characteristics of rinder-pest virus in HeLa cells. Arch. ges. Virusforsch. **14**, 27—38 (1963b).

LIESS, B., and W. PLOWRIGHT: Studies on the pathogenesis of rinderpest in experimen-tal cattle. I. Correlation of clinical signs, viraemia and virus excretion by various routes. J. Hyg. (Lond.) **62**, 81—100 (1964).

LINGARD, A.: Rep. Imp. Bact., 1905—1906; 4 (1905). Quoted by DATTA and RAJAGO-PALAN (1932).

LOWE, H. J.: Rinderpest in Tanganyika Territory. Emp. J. exp. Agric. **10**, 189—202 (1942).

LOWE, H. J., J. K. H. WILDE, R. P. LEE, and H. M. STUCHBERY: An outbreak of an aberrant type of rinderpest in Tanganyika Territory. J. comp. Path. **57**, 175—183 (1947).

LUGARD, F. D.: "The Rise of our East African Empire". Vol. I., pp. 525—528. Black-wood & Sons, Edinburgh and London (1893).

LWOFF, A., R. HORNE, and R. TOURNIER: A system of viruses. Cold Spr. Harb. Symp. quant. Biol. **27**, 51—55 (1962).

MACGREGOR, A. D.: A preliminary note on cutaneous rinderpest. Ind. J. vet. Sci. **14**, 56 (1944).

MACOWAN, K. D. S.: Research: rinderpest; in A. R. Dept. vet. Serv., Kenya, 1955; pp. 26—30 (1956).

MACOWAN, K. D. S.: A. R. Dept. vet. Serv., Kenya, 1960; pp. 4—15 (1961).

MADIN, S. H.: Tissue culture in veterinary medical research. Advanc. vet. Sci. **5**, 329—418 (1959).

MADIN, S. H., and N. B. DARBY: Established kidney cell lines of normal adult bovine and ovine origin. Proc. Soc. exp. Biol. (N.Y.) **98**, 574—576 (1958).

MALFROY, F.: La peste bovine. Étude de la maladie. Bulletin du comité d'études historiques, scientifiques de l'Afrique Occidentale Française **10**, 32—88; 275—326 (1927). Quoted by SCOTT (1964).

MALMQUIST, W. A.: Quoted by PLOWRIGHT and FERRIS (1959a).

MATUMOTO, M.: Multiplication of measles virus in cell cultures. Bact. Rev. **30**, 152—176 (1966).

MAURER, F. D.: Rinderpest XI. The survival of rinderpest virus in various mediums. Amer. J. vet. Res. **7**, 193—195 (1946).

MAURER, F. D.: Rinderpest.; in Diseases of Cattle; Ed. W. J. GIBBONS; 2nd Edit., American Veterinary Publications Inc. (1963).

MAURER, F. D., T. C. JONES, B. EASTERDAY, and D. E. DETRAY: The pathology of rinderpest. Proceedings Book, Amer. vet. med. Assoc. 92nd Ann. gen. Mtg., 1955; pp. 201—211 (1956). (See also J. Amer. vet. med. Ass. **127**, 512—514, 1955.)

MCKERCHER, P. D.: Rinderpest virus adapted to the chorio-allantoic membrane of the chick embryo. Its attenuation and use as a vaccine. Canad. J. comp. Med. **21**, 374—378 (1957).

MCKERCHER, P. D.: Plaque production by rinderpest virus in bovine kidney cultures; a preliminary report. Canad. J. comp. Med. **27**, 71—72 (1963).

MCKERCHER, P. D.: A comparison of the viruses of infectious bovine rhinotracheitis (IBR), infectious pustular vulvovaginitis (IPV) and rinderpest. Part I. Studies of antigenic relationships. Canad. J. comp. Med. **28**, 77—88 (1964a).

MCKERCHER, P. D.: A comparison of the viruses of infectious bovine rhinotracheitis (IBR), infectious pustular vaginitis (IPV) and rinderpest. Part II. Plaque assay. Canad. J. comp. Med. **28**, 113—120 (1964b).

MCKINLEY, E. B.: Experimental rinderpest in goats. Proc. Soc. exp. Biol. (N.Y.) **26**, 22 (1928).

MENON, M. S., and R. H. SAGAR: Vaccination against rinderpest with lapinised vaccine and immunity test at the Allahabad Agricultural Institute, Allahabad. Ind. vet. J. **40**, 15—22 (1963).

METTAM, R. M. W.: In A. R. Vet. Dept., Uganda, 1936; pp. 29—30 (1937).

MINETT, F. C.: The cultivation of the rinderpest virus *in vitro*. J. comp. Path. **36**, 205—216 (1923).

MITCHELL, D. T., and P. L. LEROUX: Further investigations into immunisation of cattle against rinderpest. Onderstepoort J. vet. Sci. **21**, 7—16 (1946).

MOHAN, R., and M. R. BAHL: Cutaneous eruptions of rinderpest in goats. Ind. J. vet. Sci. **23**, 39—42 (1953).

MOLINIE, J. P.: La peste bovine. Contamination à l'espèce porcine. Rec. Méd. vét. exot. **4**, 5—26 (1931).

MORCOS, Z.: Rinderpest virus and laboratory animals. Vet. Rec. **11**, 231—232 (1931).

MORNET, P., et Y. GILBERT: Les méthodes actuelles de lutte contre la peste bovine. Les Cahiers Méd. vét. **27**, 1—52 (1958).

MORNET, P., et Y. GILBERT: Bases et moyens du diagnostic de la peste bovine. Bull. Off. int. Epiz. **53**, 13—37 (1960).

MORNET, P., Y. GILBERT, J. ORUE et G. THIERY: Nouvelles recherches sur le virus-vaccin bovipestique lapinisé. Rev. Élev. **8**, 297—310 (1955).

MORNET, P., P. GORET et Y. GILBERT: Immunité croisée entre la maladie de Carré et la peste bovine. Bull. epizoot. Dis. Afr. **7**, 255—263 (1959).

MORNET, P., P. GORET, Y. GILBERT et Y. GOUEFFON: Nouvelles recherches sur l'immunisation contre la peste bovine à l'aide du virus de la maladie de Carré. C. R. Acad. Sci. (Paris) **248**, 2815—2817 (1959).

MORNET, P., P. GORET, Y. GILBERT et Y. GOUEFFON: Sur les relations croisées des charactères antigènes et immunogènes des virus de la maladie de Carré, et de la peste bovine. Etat actuel de recherches. Rev. Élev. **13**, 5—25 (1960).

MORNET, P., et M. GUERRET: Les lésions cutanées dans la peste bovine. Bull. Acad. vét. Fr. **23**, 283—285 (1950).

MORNET, P., J. ORUE et Y. GILBERT: Unicité et plasticité du virus bovipestique. A propos d'un virus naturel adapté sur petits ruminants. C. R. Acad. Sci. (Paris) **242**, 2886—2889 (1956).

MORNET, P., J. ORUE, Y. GILBERT, G. THIERY et S. MAMADOU: La ,,peste des petits ruminants" en Afrique Occidentale Francaise; ses rapports avec la peste bovine. Rev. Élev. **9**, 313—342 (1956).

MORNET, P., J. ORUE, C. LABOUCHE et P. MAINGUY: Les virus-vaccins contre la peste bovine; le virus bovipestique lapinisé: I. Revue des travaux. II. Recherches effectuées au laboratoire de Dakar. Rev. Élev. **6**, 125—166 (1953).

MOULTON, W. M., and S. S. STONE: A procedure for detecting complement-fixing antibody to rinderpest virus in heat inactivated bovine serum. Res. vet. Sci. **2**, 161—166 (1961).

MOURA, R. A.: The influence of serum on the cytopathic effect of measles virus. Arch. ges. Virusforsch. **11**, 487—492 (1962).

MUMATZEY, and KOYAMU: A study of the virulence of lapinised rinderpest virus in Shantung calves. Res. Rep. N. China Production Sci. Res. Bureau (1945). Quoted by CHENG and FISCHMAN (1949).

NAKAMURA, J.: On two serologically differentiable strains of rinderpest virus. J. Jap. Soc. vet. Sci. **10**, 367—373 (1931). Abstr. in Vet. Bull. **2**, 495 (1932).

NAKAMURA, J.: On the experimental infection with rinderpest virus in the rabbit. III. Neutralisation experiment. Jap. J. vet. Sci. **2**, 567—578 (1940). Abstr. in Vet. Bull. **11**, 761 (1941).

NAKAMURA, J.: On the experimental infection with rinderpest virus in the rabbit. V. Multiplication of the virus in the body of the infected rabbit. Jap. J. vet. Sci. **3**, 403—429 (1941). Abstr. in Vet. Bull. **22**, 187 (1952).

NAKAMURA, J.: Recherches sur l'épreuve de fixation du complément dans la peste bovine au Japon. Bull. Off. int. Epiz. **36**, 96—107 (1951).

NAKAMURA, J.: Rinderpest. Bull. Off. int. Epiz. **47**, 542—554 (1957a).

NAKAMURA, J.: Présentation de tableaux sur la peste bovine à l'Institut Nippon de science biologique le 27 Novembre, 1957. Bull. Off. int. Epiz. **47**, 555—571 (1957b).

NAKAMURA, J.: Complement-fixation reaction in rinderpest study. Guide for technique and application (71 pages). International Office of Epizootics (1958).

NAKAMURA, J.: Report to the Government of the United Arab Republic on the diagnosis of rinderpest and vaccine production. F.A.O., Rome (1962).

NAKAMURA, J.: Rinderpest: report of Japan on the technical activities performed. Bull. Off. int. Epiz. **63**, 57—72 (1965).

NAKAMURA, J., K. FUKUSHO, and S. KURODA: Rinderpest: laboratory experiments on immunisation of Chosen cattle by simultaneous inoculation with immune serum and rabbit virus. Jap. J. vet. Sci. **5**, 455—477 (1943). Abstr. in Vet. Bull. **22**, 186—187 (1952).

NAKAMURA, J., S. KISHI, K. KIUCHI, and R. REISINGER: An investigation of antibody response in cattle vaccinated with the rabbit-passaged LA rinderpest virus in Korea. Amer. J. vet. Res. **16**, 71—75 (1955).

NAKAMURA, J., S. KISHI, H. MATSUZAWA, J. KIUCHI, and T. MIYAMOTO: Inoculation experiments with attenuated strains of rinderpest virus in goats, pigs and hamsters. Bull. Nippon Inst. biol. Sci., Tokyo **2**, 1—12 (1957).

NAKAMURA, J., S. KISHI et T. MIYAMOTO: Sur les caractéristiques de la multiplication du virus lapinisé-avianisé de la peste bovine dans les embryons de poulet. Bull. Off. int. Epiz. **42**, 692—709 (1954).

NAKAMURA, J., and S. KURODA: Rinderpest: on the virulence of the attenuated rabbit virus for cattle. Jap. J. vet. Sci. **4**, 75—102 (1942). Quoted by CHENG and FISCHMAN (1949).

NAKAMURA, J., and A. J. MACLEOD: The complement fixation test and its application to the diagnosis of rinderpest. J. comp. Path. **69**, 11—19 (1959).

NAKAMURA, J., and T. MIYAMOTO: Avianisation of lapinised rinderpest virus. Amer. J. vet. Res. **14**, 307—317 (1953).

NAKAMURA, J., T. MOTOHASHI, and S. KISHI: Propagation of the lapinised-avianised strain of rinderpest virus in the culture of chicken embryo tissue. Amer. J. vet. Res. **19**, 174—180 (1958).

NAKAMURA, J., R. SATO, and T. MIYAMOTO: Infection and passage with rinderpest virus in chicken embryos through intravenous route. Jap. J. vet. Sci. **9**, 110 (1947). Quoted by NAKAMURA and MIYAMOTO (1953).

NAKAMURA, J., S. WAGATUMA, and K. FUKUSHO: On the experimental infection with rinderpest virus in the rabbit. I. Some fundamental experiments. J. Jap. Soc. vet. Sci. **17**, 185—204 (1938). Abstr. in Vet. Bull. **9**, 536 (1939).

NGUYEN-BA-LUONG, et VU-THIEN-THAI: Contribution à l'étude du virus-vaccin contre la peste bovine, souche Nakamura III. Bull. Off. int. Epiz. **50**, 559—563 (1958). Quoted by SCOTT, 1964.

Nicolas, E., et P. Rinjard: Sur la transmission de la peste bovine des bovidés au porc de race celtique. C. R. Soc. Biol. (Paris) 85, 168—170 (1921).

Nicolle, M., et Adil Bey: Études sur la peste bovine. Premier mémoire. Ann. Inst. Pasteur 13, 319—336 (1899).

Nicolle, M., et Adil Bey: Études sur la peste bovine. Deuxième mémoire. Ann. Inst. Pasteur 15, 713—733 (1901).

Nicolle, M., et Adil Bey: Études sur la peste bovine; troisième mémoire. Expériences sur la filtration du virus. Ann. Inst. Pasteur 16, 56—64 (1902).

Norrby, E. C. J.: Haemagglutination by measles virus: IV. A simple procedure for production of high potency antigen for haemagglutination-inhibition (HI) tests. Proc. Soc. exp. Biol. (N.Y.) 111, 814—818 (1962a).

Norrby, E. C. J.: Haemagglutination by measles virus: II. Properties of the haemagglutinin and of the receptors on the erythrocytes. Arch. ges. Virusforsch. 12, 164—172 (1962b).

Norrby, E. C. J., P. Magnusson, L. G. Falksveden, and M. Grönberg: Separation of measles virus components by equilibrium sedimentation in CsCl gradients. II. Studies on the large and the small haemagglutinin. Arch. ges. Virusforsch. 14, 462—473 (1964).

Oddo, F. G., R. Flaccomio, and A. Sinatra: "Giant-cell" and "strand-forming" cytopathic effect of measles virus lines conditioned by serial propagation with diluted or concentrated inoculum. Virology 13, 550—553 (1961).

Ono, S., and S. Kondo: Studies on rinderpest in deer (Cervus sika) and changes in the blood of infected animals. Author's English extract. J. Jap. Soc. vet. Sci. 2, 158—161 (1923).

Orr, W.: Rinderpest in goats imported to Malaya. J. comp. Path. 55, 185—200 (1945).

Pease, H. T.: C.V.D. Ledger Series No. I — Rinderpest (1894). Quoted by Datta and Rajagopalan (1932).

Pecaud, G.: Contribution à l'étude de la pathologie vétérinaire de la colonie du Tchad. Bull. Soc. Path. exot. 17, 196—207 (1924).

Pellegrini, D., et G. Guarini: Mise en évidence d'anticorps fixateur, dans le sérum d'animaux immunisés contre la peste bovine, avec la méthode de la fixation du complement modifée. Bull. Off. int. Epiz. 37, 233—237 (1952).

Percival, A. B.: Game and disease. J. E. Afr. Uganda Nat. Hist. Soc., No. 13, 302—315 (1918).

Periés, J. R., et C. Chany: Activité hémagglutinante et hémolytique du virus morbilleux. C. R. Acad. Sci. (Paris) 252, 2956—2957 (1960).

Pfaff, G.: Immunisation against rinderpest with special reference to the use of dried goat spleen. Onderstepoort J. vet. Sci. 11, 261—330 (1938).

Pfaff, G.: Rinderpest in buffaloes. The immunising value of dried goat spleen tissue. Onderstepoort J. vet. Sci. 15, 175—184 (1940).

Phillipe, J.: Travaux de recherches effectuées par le laboratoire de Bamako pendant le 1er semestre 1939. Bull. Serv. Zootech. Epiz. A.O.F. 1, 1—15 (1939). Quoted by Scott (1964).

Piercy, S. E., and R. D. Ferris: Establishment of a section for the tissue culture of viruses primarily in order to cultivate rinderpest virus. A. R. E. Afr. vet. Res. Org. 1954—1955, incorporating A. R. 1952 und 1953 (1957).

Piercy, S. E., G. R. Scott, and M. A. Witcomb: Studies on avianised rinderpest virus in the embryonated egg; in A. R. E. Afr. vet. Res. Org., 1956—1957, pp. 17—19 (1958).

Pinkerton, H., W. L. Smiley, and A. D. Anderson: Giant cell pneumonia with inclusions. Amer. J. Path. 21, 1—23 (1945).

Plowright, W.: Observations on the behaviour of rinderpest virus in indigenous African sheep. Brit. vet. J. 108, 450—457 (1952).

Plowright, W.: Unpublished observations (1961).

Plowright, W.: Rinderpest virus. Ann. N.Y. Acad. Sci. 101, 548—573 (1962a).

PLOWRIGHT, W.: The application of monolayer tissue culture techniques in rinderpest research. I. Introduction. Use in serological investigations and diagnosis. Bull. Off. int. Epiz. **57**, 1—23 (1962b).

PLOWRIGHT, W.: The application of monolayer tissue culture techniques in rinderpest research. II. The use of attenuated culture virus as a vaccine for cattle. Bull. Off. int. Epiz. **57**, 253—277 (1962c).

PLOWRIGHT, W.: Some properties of strains of rinderpest virus recently isolated in E. Africa. Res. vet. Sci. **4**, 96—108 (1963a).

PLOWRIGHT, W.: The production and use of culture-attenuated rinderpest vaccine. Proc. XVIIth World vet. Congr. (Hannover) **1**, 679—682 (1963b).

PLOWRIGHT, W.: The role of game animals in the epizootiology of rinderpest and malignant catarrhal fever in East Africa. Bull. epizoot. Dis. Afr. **11**, 149—162 (1963c).

PLOWRIGHT, W.: Studies on the pathogenesis of rinderpest in experimental cattle. II. Proliferation of the virus in different tissues following intranasal infection. J. Hyg. (Camb.) **62**, 267—281 (1964a).

PLOWRIGHT, W.: The growth of virulent and attenuated strains of rinderpest virus in primary calf kidney cells. Arch. ges. Virusforsch. **14**, 431—448 (1964b).

PLOWRIGHT, W.: Effect of re-vaccination on distemper antibody levels in the dog. Vet. Rec. **76**, 131 (1964c).

PLOWRIGHT, W.: Symposium "The Smallest Stowaways" — III. Rinderpest. (Presented to the BVA Ann. Congr. in Edinburgh, Sept. 1965.) Vet. Rec. **77**, 1431—1438 (1965).

PLOWRIGHT, W., J. G. CRUIKSHANK, and A. P. WATERSON: The morphology of rinderpest virus. Virology **17**, 118—122 (1962).

PLOWRIGHT, W., and R. D. FERRIS: Cytopathogenicity of rinderpest virus in tissue cultures. Nature (Lond.) **179**, 316 (1957).

PLOWRIGHT, W., and R. D. FERRIS: Studies with rinderpest virus in tissue culture. 1. Growth and cytopathogenicity. J. comp. Path. **69**, 152—172 (1959a).

PLOWRIGHT, W., and R. D. FERRIS: Studies with rinderpest virus in tissue culture. II. Pathogenicity for cattle of culture-passaged virus. J. comp. Path. **69**, 173—184 (1959b).

PLOWRIGHT, W., and R. D. FERRIS: The preparation of bovine thyroid monolayers for use in virological investigations. Res. vet. Sci. **2**, 149—152 (1961a).

PLOWRIGHT, W., and R. D. FERRIS: The serial cultivation of calf kidney cells for use in virus research. Res. vet. Sci. **2**, 387—395 (1961b).

PLOWRIGHT, W., and R. D. FERRIS: Studies with rinderpest virus in tissue culture. III. The stability of cultured virus and its use in virus neutralisation tests. Arch. ges. Virusforsch. **11**, 516—533 (1961c).

PLOWRIGHT, W., and R. D. FERRIS: Studies with rinderpest virus in tissue culture. A technique for the detection and titration of virulent virus in cattle tissues. Res. vet. Sci. **3**, 94—103 (1962a).

PLOWRIGHT, W., and R. D. FERRIS: Studies with rinderpest virus in tissue culture. The use of attenuated culture virus as a vaccine for cattle. Res. vet. Sci. **3**, 172—182 (1962b).

PLOWRIGHT, W., and K. HERNIMAN: The stability of reconstituted rinderpest culture vaccine — to be published (1967).

PLOWRIGHT, W., R. M. LAWS, and C. S. RAMPTON: Serological evidence for the susceptibility of the hippopotamus to natural infection with rinderpest virus. J. Hyg. (Lond.) **62**, 329—336 (1964).

PLOWRIGHT, W., and B. McCULLOCH: Investigations on the incidence of rinderpest virus infection in game animals of N. Tanganyika and S. Kenya 1960/63. J. Hyg. (Lond.) **65**, 343—358 (1967).

PLOWRIGHT, W., and W. P. TAYLOR: Long-term studies of the immunity in East African cattle following inoculation with rinderpest culture vaccine. Res. vet. Sci. **8**, 118—128 (1967).

POLDING, J. B., and R. M. SIMPSON: A possible immunological relationship between canine distemper and rinderpest. Vet. Rec. **69**, 582—584 (1957).

POLDING, J. B., R. M. SIMPSON, and G. R. SCOTT: Links between canine distemper and rinderpest. Vet. Rec. **71**, 643—644 (1959).

POULTON, W. F. (1914): Quoted by CARMICHAEL (1938a).

PROVOST, A.: Essais de transmission de la peste bovine par aérosols virulents. Bull. epizoot. Dis. Afr. **6**, 79—85 (1958).

PROVOST, A., et C. BORREDON: Les différents aspects du diagnostic clinique et expérimental de la peste bovine. Rev. Élev. **16**, 445—526 (1963).

PROVOST, A., C. BORREDON et R. QUEVAL: Une hypoglobulinémie essentielle des bovins d'Afrique centrale, cause d'erreur dans les enquêtes sérologiques. Rev. Élev. **18**, 385—393 (1965b).

PROVOST, A., R. QUEVAL et C. BORREDON: Quelques recherches fondamentales sur le virus bovipestique. Rev. Élev. **18**, 371—384 (1965a).

PROVOST, A., et J. M. VILLEMOT: Note sur les plasmodes multinucléés recontrés dans les cultures cellulaires infectées de virus bovipestique. Ann. Inst. Pasteur **101**, 276—280 (1961).

PROVOST, A., J. M. VILLEMOT et R. QUEYAL: La production du virus capripestique au laboratoire de Farcha. Bull. epizoot. Dis. Afr. **6**, 361—371 (1958).

RAJAGOPALAN, V. R.: Is there a relationship between the viruses of rinderpest and Doyle's disease. Ind. J. vet. Sci. **7**, 59—64 (1937).

RAPP, F., J. S. BUTEL, and C. WALLIS: Protection of measles virus by sulphate ions against thermal inactivation. J. Bact. **90**, 132—135 (1965).

RECEVEUR, P.: Réflexions sur l'épidémiologie de peste bovine en Afrique centrale. Bull. Off. int. Epiz. **37**, 536—541 (1952).

RECEVEUR, P.: Animaux sauvages dans la transmission des maladies contagieuses sauf la rage. Bull. Off. int. Epiz. **42**, 213—222 (1954).

RECEVEUR, P.: Risques de dispersion de la peste bovine par les viandes fraiches ou congelées provenant des pays contaminés. Bull. Off. int. Epiz. **48**, 148—158 (1957).

RECZKO, EVA, und K. BÖGEL: Elektronenmikroskopische Untersuchungen über das Verhalten eines vom Kalb isolierten Parainfluenza-3-Virus in Kälbernierenzellkulturen. Arch. ges. Virusforsch. **12**, 404—420 (1962).

REID, N. R.: in A. R. Dept. vet. Sci. anim. Husb., Tanganyika Territory, 1947/48 (1949).

REISSIG, M., F. L. BLACK, and J. L. MELNICK: Formation of multinucleated giant cells in measles infected cultures deprived of glutamine. Virology **2**, 836—838 (1956).

ROBERTS, G. A.: Rinderpest (peste bovina) in Brazil. J. Amer. vet. med. As. **60**, 177—185 (1921).

ROBERTSON, W. A. N.: Rinderpest in Western Australia, 1923. Service Publication (Veterinary Hygiene) No. 1; Dept. of Health, Commonwealth of Australia (1924).

ROBIN, P., et P. R. BOURDIN: Note sur l'action du sulphate de sodium, du sulphate de magnésium et du chlorure de magnésium sur le virus de la peste bovine adapté aux cultures cellulaires. Rev. Élev. **19**, 451—456 (1966).

ROBSON, J., R. M. ARNOLD, W. PLOWRIGHT, and G. R. SCOTT: The isolation from an eland of a strain of rinderpest virus attenuated for cattle. Bull. epizoot. Dis. Afr. **7**, 97—102 (1959).

ROIZMAN, B., and A. E. SCHLUEDERBERG: Virus infection of cells in mitosis. III. Cytology of mitotic and amitotic Hep-2 cells infected with measles virus. J. nat. Cancer Inst. **28**, 35—43 (1962).

ROTT, R., and W. SCHÄFER: Fine structure of subunits isolated from Newcastle disease virus (NDV). Virology **14**, 298—299 (1961).

ROTT, R., und W. SCHÄFER: Hydroxylamin-Empfindlichkeit des Newcastle-Disease-Virus (NDV). Z. Naturforsch. **17b**, 861—862 (1962).

ROTT, R., A. P. WATERSON, and I. M. REDA: Characterisation of soluble antigen derived from cells infected with Sendai and Newcastle disease viruses. Virology **21**, 663—665 (1963).

SACQUET, E., et P. TROQUEREAU: Essai de vaccination contre la peste bovine au moyen du virus capripestique dans le Nord-Est du Tchad. Rev. Élev. **5**, 45—50 (1952).

SAINTE-HILAIRE, M. A. G.: Note sur le typhus contagieux au jardin d'acclimatisation. Bull. Soc. Impériale Zoologique d'Acclimatisation, Paris **2**, 685—692 (1865). Quoted by SCOTT (1964).

SAMARTSEV, A. A., and P. N. ARBUZOV: The susceptibility of camels to glanders, rinderpest and bovine contagious pleuropneumonia. Veterinariya, Moscow **4**, 59—63 (1950). Abstr. in Vet. Bull. **15**, 396 (1945).

SANDERS, M., I. KIEM, and D. LAGUNOFF: Cultivation of viruses. Arch. Path. **56**, 148—225 (1953).

SAUNDERS, P. T., and K. K. AYYAR: An experimental study of rinderpest virus in goats in a series of 150 direct passages. Ind. J. vet. Sci. **6**, 1—86 (1936).

SCHÄFER, W., und R. ROTT: Herstellung von Virusvaccinen mit Hydroxylamin. Z. Hyg. Infekt.-Kr. **148**, 256—268 (1962).

SCHEIN, H.: Etudes sur la peste bovine. Ann. Inst. Pasteur **31**, 571—592 (1917).

SCHEIN, H.: Expériences sur la peste bovine. Bull. Soc. Path. exot. **19**, 915—928 (1926).

SCHEIN, H., and H. JACOTOT (1925): Quoted by CURASSON, 1932; 1942; reference not traced.

SCOTT, G. R.: The intradermal inoculation of lapinised rinderpest vaccine. Vet. Rec. **64**, 137—138 (1952).

SCOTT, G. R.: The virus content of the tissues of rabbits infected with rinderpest. Brit. vet. J. **110**, 152—157 (1954).

SCOTT, G. R.: Life expectancy of rinderpest virus. Bull. epizoot. Dis. Afr. **3**, 19—20 (1955a).

SCOTT, G. R.: The incidence of rinderpest in sheep and goats. Bull. epizoot. Dis. Afr. **3**, 117—118 (1955b).

SCOTT, G. R.: The risk associated with the importation of meat from countries where rinderpest control measures are still required. Bull. epizoot. Dis. Afr. **5**, 11—13 (1957).

SCOTT, G. R.: The growth parameters of Newcastle disease, Rift Valley fever and rinderpest viruses. Thesis; University of Edinburgh (1959a) — Quoted by SCOTT, 1964.

SCOTT, G. R.: A precis of the characteristics of rinderpest virus. Bull. epizoot. Dis. Afr. **7**, 173—178 (1959b).

SCOTT, G. R.: Mortality of rabbits inoculated with lapinised rinderpest virus. J. comp. Path. **69**, 148—151 (1959c).

SCOTT, G. R.: Heat inactivation of rinderpest-infected bovine tissues. Nature (Lond.) **184**, 1948—1949 (1959d).

SCOTT, G. R.: Bovine hyperimmune serum in the diagnosis of rinderpest. Vet. Rec. **74**, 409 (1962a).

SCOTT, G. R.: Optimal incubation temperature for the rinderpest agar gel double diffusion test. Bull. epizoot. Dis. Afr. **10**, 457—459 (1962b).

SCOTT, G. R.: Rinderpest. In: Advanc. vet. Sci. **9**, 113—224 (1964).

SCOTT, G. R., and R. D. BROWN: A neutralisation test for the detection of rinderpest antibodies. J. comp. Path. **68**, 308—314 (1958).

SCOTT, G. R., and R. D. BROWN: Rinderpest diagnosis with special reference to the agar gel double diffusion test. Bull. epizoot. Dis. Afr. **9**, 83—125 (1961).

SCOTT, G. R., K. M. COWAN, and R. T. ELLIOTT: Rinderpest in impala. Vet. Rec. **72**, 787—788 (1960).

SCOTT, G. R., D. E. DETRAY, and G. WHITE: A preliminary note on the susceptibility of pigs of European origin to rinderpest. Bull. Off. int. Epiz. **51**, 694—698 (1959).

SCOTT, G. R., D. E. DETRAY, and G. WHITE: Rinderpest in pigs of European origin. Amer. J. vet. Res. **23**, 452—456 (1962).

SCOTT, G. R., and J. MACDONALD: Kenya camels and rinderpest. Bull. epizoot. Dis. Afr. **10**, 495—497 (1962).

SCOTT, G. R., A. K. MACLEOD, and C. S. RAMPTON: Goats as donors of rinderpest hyperimmune serum. Vet. Rec. **75**, 1221—1222 (1963).

SCOTT, G. R., and C. S. RAMPTON: Transmission of lapinised rinderpest virus by contact between rabbits. Nature (Lond.) **192**, 289 (1961).

Scott, G. R., and C. S. Rampton: Influence of the route of exposure on the titre of rinderpest virus in rabbits. J. comp. Path. **72**, 299—302 (1962).

Scott, G. R., and M. A. Witcomb: Rinderpest virus in laboratory animals. A. R. E. Afr. vet. Res. Org., 1956—57; pp. 15—17 (1958a).

Scott, G. R., and M. A. Witcomb: Research on rinderpest vaccines: in A. R. E. Afr. vet. Res. Org., 1956—57; pp. 19—20 (1958b).

Seetharaman, C.: Contact infection studies with goat-adapted rinderpest virus. Ind. J. vet. Sci. **17**, 69—76 (1948).

Semmer, E.: Über das Rinderpest-Contagium und über Immunisierung und Schutzimpfung gegen Rinderpest. Berl. tierärztl. Wschr. **23**, 590—591 (1893).

Semmer, E. (1896): Quoted by Scott (1964).

Sharma, R. M.: Évolution et prophylaxie régionale de la peste bovine (Discussion). Bull. Off. int. Epiz. **43**, 244—245 (1965).

Sharma, R. M., and T. Ram: Scarification versus subcutaneous method of vaccination against rinderpest in goats and sheep. Ind. J. vet. Sci. **25**, 129—142 (1955).

Shilston, A. W.: The vitality of rinderpest virus outside the animal body under natural conditions. Agric. Res. Inst. Pusa. Vet. Ser. **3**, No. I, 32 pp. (1917).

Shope, R. E., and H. J. Griffiths: Rinderpest. VI. The persistence of virus in chicks hatched from infected eggs. Amer. J. vet. Res. **7**, 170—173 (1946).

Shope, R. E., H. J. Griffiths, and D. L. Jenkins: Rinderpest. I. The cultivation of rinderpest virus in the developing hen's egg. Amer. J. vet. Res. **7**, 135—141 (1946a).

Shope, R. E., F. D. Maurer, D. L. Jenkins, H. J. Griffiths, and J. A. Baker: Rinderpest. IV. Infection of the embryos and the fluids of developing hens' eggs. Amer. J. vet. Res. **7**, 152—163 (1946b).

Simmons, R.: A. R. Vet. Dept., Uganda, 1940; p. 4. Govt. Printer, Entebbe (1941).

Simon, N.: "Between the Sunlight and the Thunder". The Wildlife of Kenya. Collins, London (1962).

Simpson, S.: Vaccination against rinderpest with lapinised virus in the Gold Coast. Bull. epizoot. Dis. Afr. **2**, 6—22 (1954).

Singh, K. V., and I. F. El Cicy: Comparative cytopathology of rinderpest and bovine parainfluenza virus in cell cultures. Nature (Lond.) **211**, 314—315 (1966).

Smith, V. W.: Active immunisation of calves with tissue-cultured rinderpest vaccine. J. comp. Path. **76**, 217—224 (1966).

Stewart, D. R. M.: Rinderpest among wild animals in Kenya 1960—62. Bull. epizoot. Dis. Afr. **12**, 39—42 (1964).

Stirling, R. F.: Some experiments in rinderpest vaccination: active immunisation of Indian plains cattle by inoculation with goat-adapted virus alone in field conditions. Vet. J. **88**, 192—204 (1932).

Stirling, R. F.: Some experiments in rinderpest vaccination. (Second Report.) Vet. J. **89**, 290—306 (1933).

Stone, S. S.: Multiple components of rinderpest virus as determined by the precipitin reaction in agar gel. Virology **11**, 638—640 (1960).

Stone, S. S., and P. D. DeLay: The inactivation of rinderpest virus by β-propriolactone and its effect on homologous complement-fixing and neutralising antibodies. J. Immunol. **87**, 464—467 (1961).

Stone, S. S., and W. M. Moulton: A rapid serologic test for rinderpest. Amer. J. vet. Res. **22**, 18—22 (1961).

Strickland, K. L.: Vaccination of calves against rinderpest. Vet. Rec. **74**, 630—631 (1962).

Syntin, S. P.: The infection of wild swine by the horned cattle plague. Vestrik Sovremennoi Veterinariya No. 11, 348—349 (1928). Quoted by Scott (1964).

Tajima, M., T. Ushijima, S. Kishi, and J. Nakamura: Electron microscopy of cytoplasmic inclusion bodies in cells infected with rinderpest virus. Virology **31**, 92—100 (1967).

Takematsu, M., and T. Morimoto: Studies on tissue culture with rinderpest virus. Jap. J. vet. Sci. **16**, 185 (1954). Abstr. in Vet. Bull. **26**, 380 (1956).

Taylor, W. P.: unpublished observations (1964).

Taylor, W. P.: The susceptibility of the one-humped camel *(Camelus dromedarius)* to infection with rinderpest virus — to be published (1967).

Taylor, W. P., and W. Plowright: Studies on the pathogenesis of rinderpest in experimental cattle. III. Proliferation of an attenuated strain in various tissues following subcutaneous inoculation. J. Hyg. (Lond.) **63**, 263—275 (1965).

Taylor, W. P., W. Plowright, R. Pillinger, C. S. Rampton, and R. F. Staple: Studies on the pathogenesis of rinderpest in experimental cattle. IV. Proliferation of the virus following contact infection. J. Hyg. (Lond.) **63**, 497—506 (1965).

Theiler, A.: Rinderpest in Südafrika. Schweiz. Arch. Tierheilk. **39**, 49—62 (1897a).

Theiler, A.: Experimentaluntersuchungen über Rinderpest. Schweiz. Arch. Tierheilk. **39**, 193—213 (1897b).

Thiery, G.: Hématologie, histopathologie et histochimie de la peste bovine. Rev. Élev. **9**, 117—140 (1956).

Thomas, A. D., and N. R. Reid: Rinderpest in game; a description of an outbreak and an attempt at limiting its spread by means of a bush fence. Onderstepoort J. vet. Sci. anim. Indust. **20**, 7—23 (1944).

Thome, M.: *in* Laboratoire de Farcha; Rapport annuel, 1964 (1965).

Todd, C., and R. G. White: Experiments on cattle plague. Cairo Govt. Press., pp. 1—133 (1914).

Tokuda, G., K. Fukusho, T. Morimoto, and M. Watanabe: Studies on rinderpest virus in bovine leucocyte culture. I. Cultivation of leucocytes and appearance of inclusions in infected cells. Nat. Inst. Anim. Hlth. Quart. **2**, 189—200 (1962).

Tokuda, G., K. Fukusho, T. Morimoto, and M. Watanabe: Studies on rinderpest virus in bovine leucocyte culture. II. Susceptibility of leucocyte culture to the virus. Nat. Inst. anim. Hlth. Quart. **3**, 55—63 (1963).

Topacio, T.: The use of goat virus in simultaneous inoculation against rinderpest (preliminary report). Philipp. agric. Rev. **19**, 297—309 (1926).

Van Saceghem, R.: La peau, voie de pénétration pour le virus de la peste bovine. C. R. Soc. Biol. (Paris) **88**, 142—143 (1923).

Verma, S. K.: Évolution et prophylaxie régionale de la peste bovine (Discussion). Bull. Off. int. Epiz. **43**, 245 (1965).

Villemot, J. M., A. Provost et P. Goret: Nouvelles recherches sur l'immunisation croisée: maladie de Carré — peste bovine. Rev. Élev. **14**, 233—244 (1961).

Vittoz, R.: Report of the Director on the work, accomplishments and administrative activities of the Office International des Epizooties for the period May, 1961 to May, 1962. Bull. Off. int. Epiz. **58**, 1131—1250 (1962).

Vittoz, R.: Report of the Director on the Scientific and Technical Activities of the Office International des Epizooties from May, 1962 to May, 1963, p. 5. Office International des Epizooties, Paris (1963),

Waddington, F. G.: An experiment to test infectivity of cattle which are reacting to KAG virus. Vet. Rec. **57**, 479—481 (1945).

Walker, J.: Rinderpest research in Kenya. Bull. No. 8A; Dept. of Agriculture: Colony and Protectorate of Kenya; pp. 6—7 (1929).

Walker, R. V. L.: Rinderpest studies. Attenuation of rabbit-adapted strain of rinderpest virus. Canad. J. comp. Med. **11**, 11—16 (1947).

Walker, R. V. L., J. A. Baker, and D. L. Jenkins: Rinderpest. II. Certain immunity reactions. Amer. J. vet. Res. **7**, 142—144 (1946a).

Walker, R. V. L., H. J. Griffiths, R. E. Shope, F. D. Maurer, and D. L. Jenkins: Rinderpest. III. Immunisation experiments with inactivated bovine tissue vaccines. Amer. J. vet. Res. **7**, 145—151 (1946b).

Ward, A. R., F. W. Wood, and W. H. Boynton: Experiments on the transmission of rinderpest. Philipp. J. Sci. **9**, 49—78 (1914).

Warren, J. L.: The relationships of the viruses of measles, canine distemper and rinderpest. Advanc. Virus Res. **7**, 27—60 (1960).

Waterson, A. P.: Two kinds of myxovirus. Nature (Lond.) **193**, 1163—1164 (1962).

Waterson, A. P.: Measles virus. Arch. ges. Virusforsch. **16**, 57—80 (1965).

WATERSON, A. P., J. G. CRUIKSHANK, G. D. LAURENCE, and A. D. KANAREK: The nature of measles virus. Virology 15, 379—382 (1961).

WATERSON, A. P., R. ROTT, and G. RUCKLE-ENDERS: The components of measles virus and their relation to rinderpest and distemper. Z. Naturforsch. 18b, 377—384 (1963).

WELLER, T. H., and A. H. COONS: Fluorescent antibody studies with agents of varicella and herpes zoster propagated *in vitro*. Proc. Soc. exp. Biol. (N.Y.) 86, 789—794 (1954).

WESTON, E. A.: Rinderpest in Australia. J. Amer. vet. med. Ass. 66, 337—350 (1924).

WHITE, G.: The agar diffusion precipitation test for rinderpest: *in* A. R. E. Afr. vet. Res. Org., 1956—57, pp. 22—23 (1958a).

WHITE, G.: A specific diffusible antigen of rinderpest virus demonstrated by the agar double-diffusion precipitation reaction. Nature (Lond.) 181, 1409 (1958b).

WHITE, G., and K. M. COWAN: Separation of the soluble antigens and infectious particles of rinderpest and canine distemper. Virology 16, 209—211 (1962).

WHITE, G., and G. R. SCOTT: An indirect gel diffusion precipitation test for the detection of rinderpest antibody in convalescent cattle. Res. vet. Sci. 1, 226—229 (1960).

WHITE, G., R. M. SIMPSON, and G. R. SCOTT: An antigenic relationship between the viruses of bovine rinderpest and canine distemper. Immunology 4, 203—205 (1961).

WILDE, J. K. H.: Rinderpest in some African wild mammals. J. comp. Path. 58, 64—72 (1948).

WILDE, J. K. H.: In: A. R. Vet. Dept. Tanganyika, 1948; pp. 11—17 (1949).

WILDE, J. K. H.: The game animal factor in the control of rinderpest in tropical Africa. Proc. XVth. Int. vet. Congr., Stockholm; 1, 283—287 (1953).

WILDE, J. K. H., and G. R. SCOTT: Rinderpest interference with caprinised rinderpest virus. J. comp. Path. 71, 222—227 (1961).

WITCOMB, M. A., S. E. PIERCY, and G. R. SCOTT: Mortality of fowl embryos inoculated with avianised strains of rinderpest virus. Res. vet. Sci. 3, 111—117 (1962).

WOOLLEY, P. G.: Rinderpest. Philipp. J. Sci. 1, 577—616 (1906).

ZWART, D., and I. MACADAM: Transmission of rinderpest by contact from cattle to sheep and goats. Res. vet. Sci. 8, 37—47 (1967a).

ZWART, D., and I. MACADAM: Observations on rinderpest in sheep and goats and transmission to cattle. Res. vet. Sci. 8, 53—57 (1967b).

ZWART, D., and L. W. ROWE: The occurrence of rinderpest antibodies in the sera of sheep and goats in Northern Nigeria. Res. vet. Sci. 7, 504—511 (1966).

Lumpy Skin Disease Virus

By

K. E. Weiss

Department of Agriculture, Veterinary Research Institute, Onderstepoort,
Republic of South Africa

With 10 Figures

Table of Contents

I. Introduction and History

Lumpy skin disease is an acute, subacute or inapparent viral disease of cattle, characterised by fever and the sudden appearance of firm circumscribed skin nodules which usually undergo necrosis. Similar lesions may be present in the skeletal muscles and the mucosae of the digestive and respiratory tracts. A subcutaneous oedema of the limbs and ventral parts of the body and a generalized lymphadenitis are also characteristic of the disease.

The virus is a member of the pox group and has many characteristics in common with the vaccinia-variola subgroup.

Although the description which follows deals essentially with the above-mentioned virus, it is considered expedient at this stage to draw attention to the fact that a disease, which closely resembles lumpy skin disease clinically, is caused by a completely unrelated virus presumably of the herpes virus group. In a review article on lumpy skin disease, the clinical manifestations and growth characteristics of this, so-called, "Allerton" virus have been described (WEISS, 1963). Subsequently, a description of the lesions caused by a related agent, designated mammilitis virus, has been given by MARTIN, MARTIN, HAY and LAUDER (1966).

In 1929, long before the etiology of lumpy skin disease was known, a skin disease of cattle, called "pseudo-urticaria" was noticed in the territory then known as Northern Rhodesia. This disease was presumably lumpy skin disease, but at the time the lesions were thought to be caused by the bites of insects (MAC-DONALD, 1931; MORRIS, 1931). The disease continued to make its appearance in subsequent years and was also ascribed to plant poisoning (LE ROUX, 1945).

Lumpy skin disease was first recognised as an infectious malady by VON BACKSTRÖM (1945), when an outbreak occurred in Ngamiland during 1943. Towards the end of 1944 the disease made its first appearance in the Transvaal (THOMAS and MARÉ, 1945) and then spread rapidly throughout South Africa during the ensuing years, despite enforced control measures. Southern Rhodesia became infected in 1945 (HUSTON, 1945) and by 1947 the disease had become firmly established and enzootic in South Africa and had also been reported from Swaziland, Basutoland and Portuguese East Africa and subsequently from Madagascar, Tanganyika and the Belgian Congo (DIESEL, 1949; DE SOUSA DIAS and LIMPO-SERRA, 1956; HAIG, 1957). The first outbreak of the disease in Kenya occurred towards the end of 1957 (MACOWEN, 1959).

In South Africa, sporadic outbreaks of lumpy skin disease recurred throughout the ensuing years, with the exception of 1953/54, 1957, 1962 and 1967, when the disease assumed epizootic proportions (HAIG, 1957; WEISS, 1962, 1967).

Although VON BACKSTRÖM (1945) concluded that lumpy skin disease was infectious, THOMAS, ROBINSON and ALEXANDER (1945) were the first to demonstrate the transmissibility of the infectious agent by the subinoculation of suspensions of skin nodules. In 1948, VAN DEN ENDE, ALEXANDER, DON and KIPPS isolated a virus in embryonated eggs, which they believed to be the cause of lumpy skin disease, but which was subsequently shown to have no etiological relationship to the disease (HAIG, 1957; ALEXANDER, PLOWRIGHT and HAIG, 1957). The virus causing lumpy skin disease was first isolated in tissue culture by ALEXANDER et al. (1957). Subsequently it was recovered on numerous occasions from the skin lesions of infected cattle in South Africa as well as in Kenya and there is no doubt that the Group III viruses (prototype Neethling), described by ALEXANDER et al. (1957), is the cause of lumpy skin disease (ALEXANDER and WEISS, 1959; PRYDIE and COACKLEY, 1959).

II. Classification of the Virus

Available evidence indicates that the virus of lumpy skin disease is a member of the pox group. The following observations are given in support of this view:

a) The cytopathic changes observed in infected tissue culture cells and particularly the intracytoplasmic inclusion bodies resemble those produced by other pox viruses (ALEXANDER et al., 1957; PRYDIE and COACKLEY, 1959; PLOWRIGHT and WITCOMB, 1959).

b) The inoculation of tissue culture virus onto the chorioallantoic membranes of developing chick embryos causes the development of macroscopic "pocks", although the appearance of the lesions are not quite similar to those caused by other pox viruses (ALEXANDER et al., 1957; VAN ROOYEN and MUNZ, 1967).

c) On intradermal inoculation of rabbits the virus produces local skin lesions followed by generalization within four days (ALEXANDER et al., 1957).

d) It is a DNA virus similar to the other members of the pox group (WEISS and BROEKMAN, 1965).

e) Lumpy skin disease virus is morphologically similar to vaccinia virus (MUNZ and OWEN, 1966).

f) CAPSTICK, PRYDIE, COACKLEY and BURDIN (1959) and CAPSTICK and COACKLEY (1961), demonstrated an antigenic relationship between the viruses of lumpy skin disease and sheep pox.

III. Properties of the Virus

A. Morphology

After purification of tissue culture propagated virus, MUNZ and OWEN (1966) examined negatively stained preparations in the electron microscope. In preparations stained at pH 6.5, they observed virus particles morphologically not unlike the M-forms of vaccinia virus described by WESTWOOD, HARRIS, ZWARTOUW, TITMUSS and APPLEYARD (1964) or alternatively the type 1 vaccinia virus particles of NAGINGTON and HORNE (1962) and MÜLLER and PETERS (1963). These lumpy skin disease virus particles, depicted in Fig. 1, consisted of a complex interwoven network of strands each with an approximate width of 70 to 90 Å and presenting an irregular surface structure. A regular arrangement of "teeth" along the margins of some particles was evident.

Preparations negatively stained at pH 8.5 showed a predominance of particles with multilayered capsules as shown in Fig. 2. These particles did not show the thread-like surface structure but consisted of a homogeneous granular mass enveloped by a capsule approximately 280 Å thick. The capsule appeared

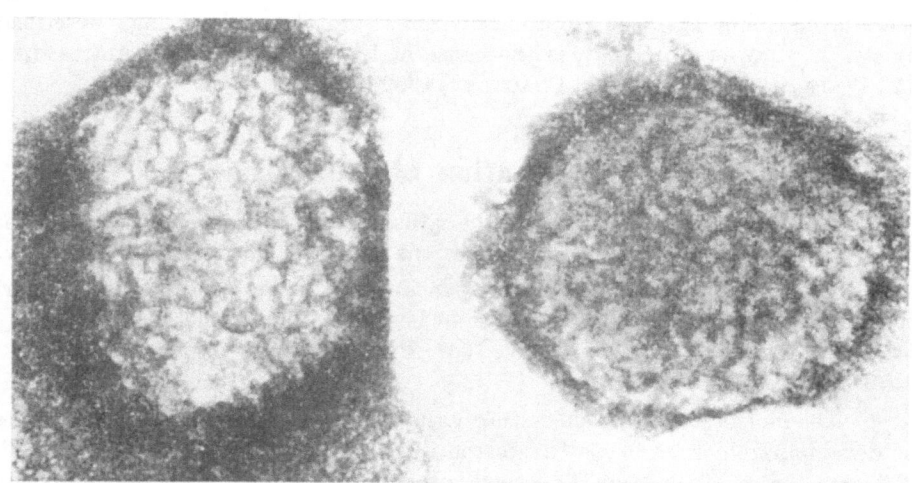

Fig. 1. Electron micrograph of lumpy skin disease virus particles stained with phosphotungstic acid at pH 6.5. Magnification, 200,000 ×. (Reprinted with kind permission of the Chief: Veterinary Research Institute, Onderstepoort.)

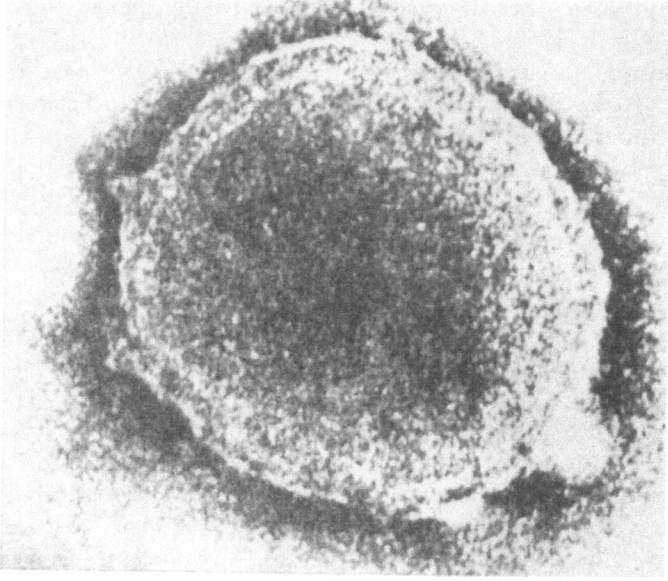

Fig. 2. Electron micrograph of lumpy skin disease virus particles stained with phosphotungstic acid at pH 8.5. Magnification, 200,000 ×. (Reprinted with kind permission of the Chief: Veterinary Research Institute, Onderstepoort.)

to have an irregular outer membrane, a thicker central part and a distinct inner membrane. These particles, in turn, resembled the C-forms of vaccinia virus described by Westwood et al. (1964), or the type 2 vaccinia virus particles of Nagington and Horne (1962) and Müller and Peters (1963).

Both C- and M-forms were observed in preparations stained at pH 7.0, an observation in agreement with that made with other pox viruses. The virus particles measured approximately 3500 Å in length and 3000 Å in width, with an axis ratio in the region of 1.2 (Munz and Owen, 1966).

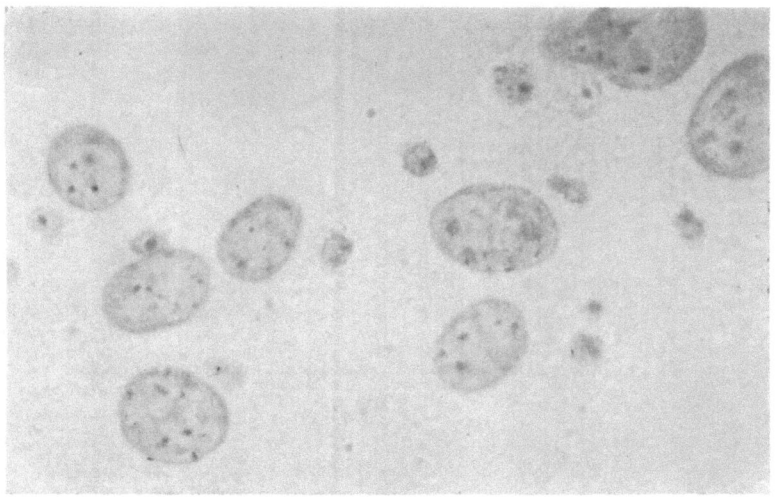

Fig. 3. Inclusion bodies of lumpy skin disease virus in tissue culture stained according to the Schiff procedure. Magnification 400 ×.

B. Physico-chemical Structure

The nucleic acid incorporated in the structure of lumpy skin disease virus is essentially DNA. In tissue cultures of lamb kidney cells infected with the "Neethling" strain of virus, Weiss and Broekman (1965) showed that the intracytoplasmic inclusion bodies, produced during multiplication of the virus, exhibit the histochemical staining reaction for DNA. With acridine orange staining, the inclusion bodies were observed as bright green, round or irregularly shaped masses in the cytoplasm, which appeared brick red with fluorescence microscopy. Most of the affected cells contained single inclusions close to the nucleus, but sometimes a number of bodies of varying sizes were seen in the same cell. With the Schiff, the Schiff methylene blue and fluorescent Feulgen staining procedures, the inclusion bodies were Feulgen positive. Photomicrographs of infected tissue culture preparations stained with the various histochemical methods are reproduced in Figs. 3 and 4.

By means of the fluorescent antibody technique, Weiss and Broekman (1965) have shown that the inclusion bodies in infected tissue culture cells consist of viral antigen. This is illustrated in Fig. 5.

The effect of metabolic inhibitors was investigated by Weiss (1966) who found that multiplication of lumpy skin disease virus in tissue cultures of lamb

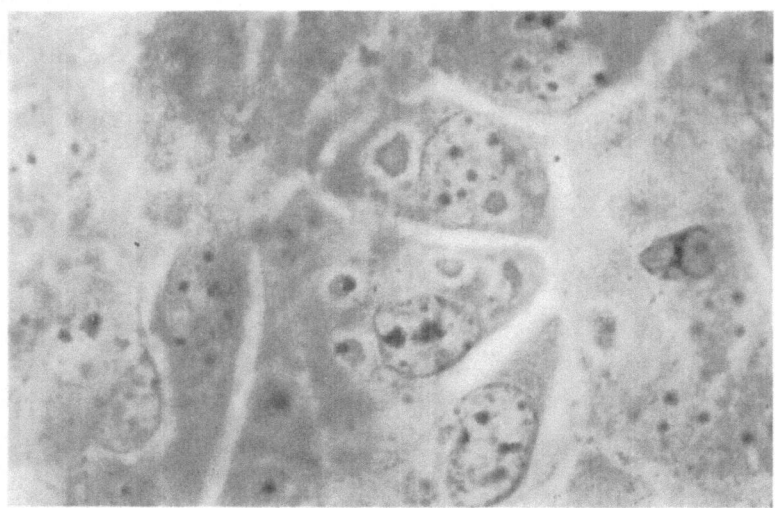

Fig. 4. Inclusion bodies of lumpy skin disease virus in tissue culture stained according to the Schiff methylene blue method. Magnification 400 ×.

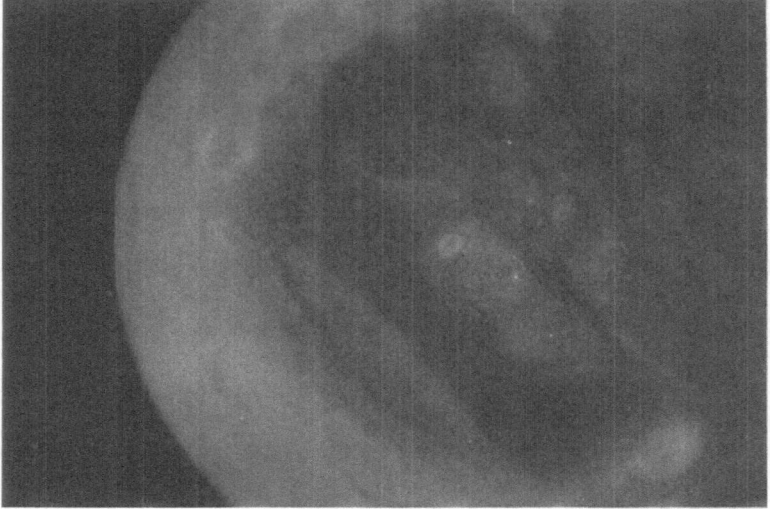

Fig. 5. Inclusion bodies of lumpy skin disease virus in tissue culture stained with fluorescent antibody. Magnification 400 ×. (Figs. 3 to 5 reprinted with kind permission of the Chief: Veterinary Research Institute, Onderstepoort.)

kidney cells was completely inhibited by 5-bromodeoxyuridine (BUDR), which has been shown to be incorporated in the viral DNA in the place of thymidine, thus rendering the virus particles non-infectious (EASTERBROOK and DAVERN, 1965). This is additional evidence in support of the statement that lumpy skin disease is a DNA-containing virus similar to the other known pox viruses.

Certain physical characteristics, such as the buoyant density of the virus, has not yet been determined.

C. Antigenic Structure

Lumpy skin disease virus has been isolated from a large number of field specimens originating from widely separated outbreaks of the disease in South Africa, Kenya and Malawi (WEISS. 1962; PRYDIE and COACKLEY, 1959; WEISS, 1966). All the virus isolations, belonging to the pox group, were antigenically similar and showed complete reciprocal cross-neutralization with the "Neethling" prototype strain. Complete cross-immunity between the South African "Neethling" virus and the Londiani strain, isolated in Kenya, has also been demonstrated (PRYDIE and COACKLEY, 1959). The available evidence suggests that there is only one immunological type of virus responsible for true lumpy skin disease.

CAPSTICK (1959) investigated the antigenic relationship of lumpy skin disease virus to sheep pox. He reported that cattle, inoculated intradermally with the Isiolo strain of virus isolated from sheep showing lesions resembling sheep pox, developed lesions indistinguishable from those of lumpy skin disease and that these animals were subsequently immune to challenge with lumpy skin disease virus.

According to CAPSTICK, who is cited by ANDREWES (1964), the Isiolo strain of sheep pox virus shows a closer relationship to goat pox virus. True sheep pox virus apparently does not protect cattle as well against lumpy skin disease.

The immunological relationship to sheep pox was further substantiated by CAPSTICK and COACKLEY (1961), who showed that cattle could be protected against lumpy skin disease by vaccination with a strain of sheep pox virus grown in tissue culture.

In serum-virus neutralization tests in tissue culture, WEISS (1959), failed to demonstrate significant in vitro neutralization of vaccinia virus by an antiserum which neutralized at least 3.0 logs of Neethling virus. However, since a sensitive plaque assay method was not employed, these results cannot be regarded as conclusive.

D. Resistance to Physical and Chemical Agents

Lumpy skin disease virus is remarkably stable between pH 6.6 and 8.6 and will show no significant reduction in titre after exposure for 5 days at 37°C within the pH range mentioned above. The virus is readily inactivated by the detergent sodium-dodecyl-sulphate, and is chloroform and ether sensitive, suggesting that a lipid is incorporated in the structure of the virus (PLOWRIGHT and FERRIS, 1959; WEISS, 1959).

In the skin lesions of infected animals, the virus can persist for at least 33 days even though the necrotic portions of skin have become completely dried out prior to sequestration. It has also been shown that the virus remained viable for 18 days in the lesions and superficial epidermal scrapings from such lesions in air-dried portions of hide kept at room temperature (ALEXANDER and WEISS, 1959). The virus was also recovered from intact skin nodules kept at −80°C for 10 years (WEISS, 1967), and from infected tissue culture fluid kept at 4°C for 6 months (ALEXANDER and WEISS, 1959). Virus in tissue culture fluid has remained viable for at least 10 years under dry-ice refrigeration (WEISS, 1967). This is contrary to the report by HAIG (1957) that prolonged storage of preparations of lumpy skin disease virus under dry-ice refrigeration reduced the infectivity.

E. Cultivation

Lumpy skin disease virus can be propagated in a large variety of animal cells grown in tissue culture. Besides lamb and calf kidney tissue cultures, the virus multiplies and produces cytopathic changes in cultures of lamb and calf testes, the sheep kidney cell strain of MADIN and DARBY (1959), lamb and calf adrenal and thyroid cultures, foetal lamb and calf muscle, sheep embryonic kidney and lung, rabbit foetal kidney and skin, chicken embryo fibroblasts, in a line of adult vervet monkey kidney cells (AVK 58)[1] and in the line of baby hamster kidney cells (BHK/21) (ALEXANDER et al., 1957; WEISS, 1959, 1964; PRYDIE and COACKLEY, 1959; MUNZ, 1965; VAN ROOYEN and WEISS, 1966).

With primary isolation of the virus in lamb kidney monolayers maintained in Hanks' balanced salt solution with 0.5% lactalbumin hydrolysate and 5.0% bovine serum, cytopathic changes were evident after an average of 11 days and thereafter progressed extremely slowly. With serial passage of the virus in cultures, the changes became visible on the 3rd day. Cytopathic changes initially appeared as discreet foci or clumps of rounded and more refractile cells. These foci gradually enlarged involving cells in the immediate surrounding, while new foci appeared in the rest of the cell sheet (ALEXANDER et al., 1957; PRYDIE and COACKLEY, 1959; DE LANGE, 1959). Although some of the affected cells eventually detached leaving irregular holes in the cell sheet, the majority usually adhered to the glass surface for a considerable time. In cultures infected with dilute suspensions of the virus, these foci or micro-plaques could be counted and constituted a sensitive method of virus assay (WEISS, 1962, 1966). Similar foci have been observed in cultures of foetal muscle cells infected with Neethling virus (MUNZ, 1967).

The rate of appearance and the progress of cytopathic changes in lamb kidney monolayers were found to be markedly increased by raising the concentration of lactalbumin hydrolysate in the medium to 2.0%. With primary isolation of the virus from field specimens, cytopathic changes were observed as early as the 3rd day and with tissue culture adapted virus these changes were apparent in 24 hours and complete in 48 to 72 hours (WEISS and GEYER, 1959). A marked increase in the rate of cytopathogenesis has also been observed in lamb kidney cultures maintained in Hanks' medium, without lactalbumin hydrolysate and serum, in which the electrolyte concentration was reduced to half the physiological strength (WEISS and WEYLAND, 1960).

---

[1] Kindly supplied by the South African Poliomyelitis Research Foundation, Johannesburg.

In infected tissue culture cells the virus is mostly cell-bound and can be released by disruption of cells with low-frequency supersonic waves. Centrifuged supernatant fluid from infected lamb kidney cultures usually contain 5.5 to 6.0 logs of virus, whereas washed supersonic-treated cells yield 7.5 to 8.0 logs of virus (WEISS, 1959).

Microscopic examination of infected monolayers, stained with haematoxylin-eosin or haematoxylin-phloxin, showed that cytopathic changes are accompanied by the development of intracytoplasmic inclusion bodies similar to those described by THOMAS and MARÉ (1945) in histological sections of the skin lesions of animals suffering from lumpy skin disease (DE LANGE, 1959; PRYDIE and COACKLEY, 1959; BURDIN, 1959). In tissue cultures, these inclusions at first appear as small

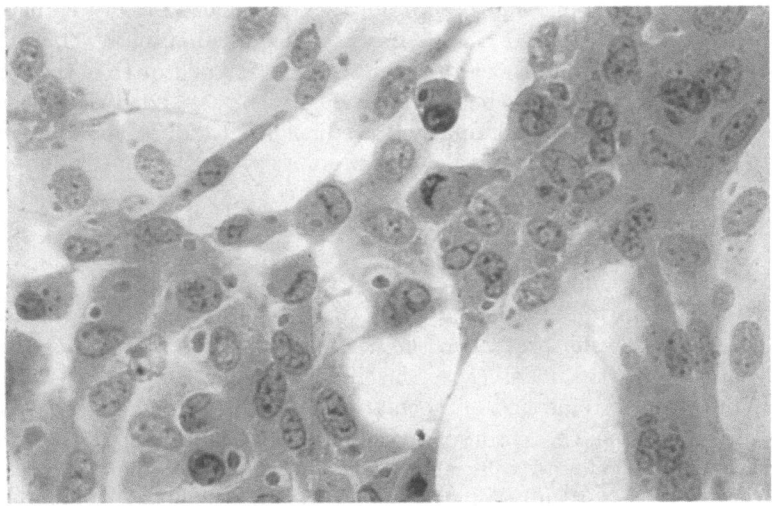

Fig. 6. Inclusion bodies of lumpy skin disease virus in infected lamb kidney tissue cultures. Haematoxylin-eosin staining. × 140. (Reprinted with kind permission of the Chief: Veterinary Research Institute, Onderstepoort.)

round basophilic bodies surrounded by a halo. As they increase in size, they become more acidophilic and some inclusions appear to have basophilic "inner bodies", which have been shown to consist of cytoplasmic RNA by histochemical staining methods (WEISS and BROEKMAN, 1965). Some inclusion bodies are round and others have an irregular outline and show small protuberances at their margins. Cells may contain one to several inclusion bodies of varying sizes. These inclusions resemble those of sheep pox (PLOWRIGHT and WITCOMB, 1959). Affected cells become rounded and shrunken, the cytoplasm becomes intensely eosinophilic and the nuclei show degenerative changes consisting of margination of chromatin, juxtaposition of the nucleoli to the nuclear membrane, and eventual pyknosis and distortion (DE LANGE, 1959; PRYDIE and COACKLEY, 1959). The inclusion bodies are shown in Fig. 6.

By means of fluorescent antibody staining, WEISS and BROEKMAN (1965) have conclusively demonstrated that the cytoplasm is the site of virus multiplication and that the inclusion bodies consist of viral antigen. In growth studies it was found that the rate of multiplication of the virus coincided with the appearance and increase in the number of inclusion bodies (PLOWRIGHT and WITCOMB, 1959).

The virus of lumpy skin disease can also be propagated in the chorio-allantoic membrane (CAM) of embryonated hens' eggs. HAIG (1957) was the first to report cultivation of the virus in the CAM but he failed to maintain the virus by serial passage in eggs. ALEXANDER et al. (1957) also found that the virus produced macroscopic "pocks" in the CAM of embryonated eggs. Subsequently VAN ROOYEN, KÜMM, WEISS and ALEXANDER (1959) succeeded in adapting and passaging Neethling virus in the CAM. This strain was subsequently passaged in eggs for at least 45 generations and was found to be attenuated for cattle after 20 serial passages (WEISS, 1960). VAN ROOYEN et al. (1959) found that the highest yield of virus could be obtained by inoculating 5-day embryonated eggs and harvesting the CAMs after 6 days further incubation at 33.5°C. Apart from congestion of the membranes and slight retardation of growth of the embryos, no macroscopic lesions were seen in the CAMs under these conditions. Evidence of virus multiplication was obtained by seeding lamb kidney monolayers with membrane suspensions in which the average infectivity was approximately $10^{4,5}$ $TCID_{50}$.

In a comprehensive study of the factors influencing the growth of the virus in embryonated eggs and the development of lesions in the CAMs, VAN ROOYEN and MUNZ (1965) found that optimal multiplication occurred in 5- to 7-day embryonated eggs incubated at 33.5 to 35°C for 4 to 6 days — the highest concentrations of virus being present in the CAMs irrespective of whether the virus was inoculated onto the membranes, into the allantoic sac, into the yolk sac or by means of the stab method of inoculation. The second highest concentration of virus was found in the embryos, with negligible amounts in the allantoic fluid and yolk. They found that under certain conditions macroscopic lesions developed in the CAMs of embryonated eggs as illustrated in Fig. 7. Small discreet whitish plaques or stripes developed most readily in the CAMs of 7- to 9-day embryonated eggs inoculated onto the membranes and incubated at 33.5 to 35°C for 7 days. Lesions failed to develop in the membranes of 5-day embryos which were found to be optimal for virus multiplication. An incubation temperature of 37°C and other routes of inoculation were unfavourable for the development of lesions. In electron micrographs of lesions in infected CAMs, these authors demonstrated particles of lumpy skin disease virus in the cytoplasm of ectodermal cells, thus proving that the lesions are associated with virus multiplication. Although the development of these lesions could be inhibited by immune serum, they concluded that the presence of lesions under certain conditions was not associated with, or an indication of optimal virus reproduction.

F. Pathogenicity and Pathogenesis

ALEXANDER et al. (1957) and MUNZ (1965) showed that rabbits inoculated intradermally with lumpy skin disease virus grown in tissue culture, developed local erythematous swellings at the site of inoculation. According to ALEXANDER et al. (1957) the local lesions were followed by generalization within 4 days, but these observations have not been confirmed subsequently (WEISS, 1967).

The natural hosts of the virus are undoubtedly cattle of which all breeds and both sexes appear to be equally susceptible. In South Africa, the disease has been observed only in cattle under natural conditions, but sheep and goats infected experimentally by intradermal inoculation develop a local erythematous swelling which disappears rapidly (ALEXANDER and WEISS, 1959). In Kenya, the disease has affected indigenous Zebu cattle as well as exotic breeds (MacOWEN, 1959). BURDIN and PRYDIE (1959) suggested that the first outbreak of lumpy skin disease in Kenya might have been caused by introduction of sheep showing a pox-like disease. In support of this contention, CAPSTICK (1959) showed that sheep and goats injected intradermally with the Londiani strain of lumpy skin disease virus developed lesions indistinguishable from those in cattle. This strain

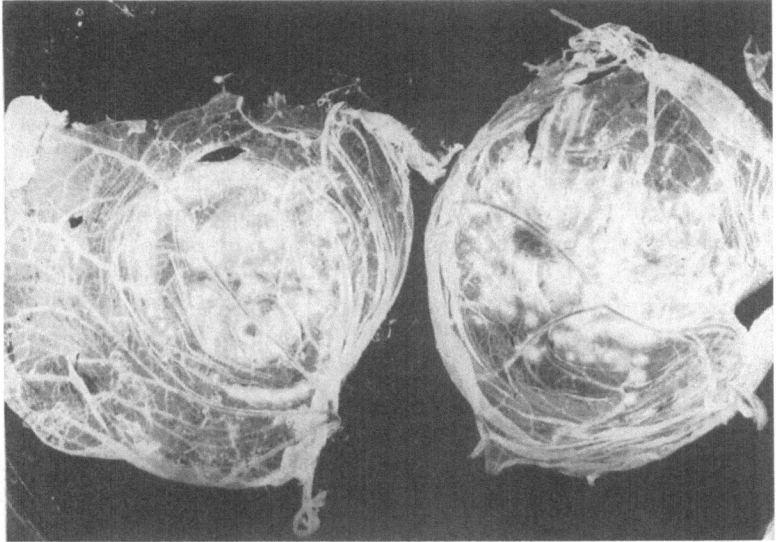

Fig. 7. Lesions produced by lumpy skin disease virus in the chorio-allantoic membranes of 7-day embryonated eggs incubated at 35°C for 7 days. (Reprinted with kind permission of the Chief: Veterinary Research Institute, Onderstepoort.)

of virus was successfully passaged in sheep but not in goats. However, during serial passage in sheep the virus apparently became contaminated with a virus resembling sheep pox, which is present in Kenya. The Isiolo strain of virus, isolated from sheep, was injected intradermally in cattle and produced lesions similar to lumpy skin disease. From these observations, CAPSTICK (1959) suggested that sheep might possibly act as carriers of lumpy skin disease virus. It has not been possible to confirm these observations and conclusions in South Africa, which is free from sheep pox.

Although cattle are the natural hosts of lumpy skin disease virus, a large percentage of animals, irrespective of breed, appear to have a natural resistance and do not develop overt symptoms of disease following natural infection. According to HAIG (1957), the morbidity rate on different farms in South Africa varied between 5 and 45%; VON BAKSTRÖM (1945) reported morbidity rates of between 50 and 100% in Ngamiland and in the outbreak in Kenya only a limited number of animals were involved; for instance, on 20 farms only single cases were observed; on 6 farms only 2 cases; on 3 farms only 5 cases and on 21 farms several cases appeared among the herds (MacOwen, 1959; Burdin and Prydie, 1959). In experimentally produced lumpy skin disease only 40 to 50% of inoculated animals developed generalized skin lesions. The remaining animals either developed localized and circumscribed painful swellings of varying size at the site of inoculation or showed no detectable reaction at all (Weiss, 1959). Capstick (1959) also reported that generalization occurred in only 3 out of 56 animals inoculated with the virus.

Following subcutaneous or intradermal inoculation of cattle with virus suspensions, a local firm painful swelling may develop at the site of inoculation after an incubation period of 4 to 7 days. The swelling may be up to 20 cms or more in diameter, circumscribed or diffuse and involving both the skin and subcutaneous tissue and sometimes the underlying musculature. The regional lymph glands are enlarged. Following the development of the local reaction, a generalized eruption of skin nodules may occur within 48 hours of the first rise of body temperature (Alexander et al., 1957; Alexander and Weiss, 1959). According to Capstick (1959) generalization usually occurs 7 to 19 days after inoculation of the virus. The local swelling may persist for longer than 6 weeks but regression with necrosis of the skin usually occurs.

In the infected animal, virus is present in the skin nodules, muscles, blood, spleen and saliva (Thomas, Robinson and Alexander, 1945; Adelaar and Neitz, 1948; Haig, 1957; Capstick, 1959). In experimentally infected animals, Alexander and Weiss (1959) demonstrated the presence of virus in blood for a period of 4 days following the appearance of fever and generalized skin lesions; in saliva for 11 days; in semen of bulls for 22 days and in the skin nodules themselves for a period of 33 days following their first appearance. At this stage the skin lesions were transformed into detachable dry necrotic skin sequesters. Virus was also recovered from the apparently normal skin of infected animals on the 7th and 12th day of the disease and from the saliva and semen of bulls undergoing an inapparent reaction. The nasal discharge of animals with lesions in the nasal mucous membrane was found to be infective, but virus could not be demonstrated in the urine and faeces.

Lumpy skin disease has not yet been reported as a natural infection in game animals. However, Young, Basson and Weiss (1966) showed that the giraffe [Giraffa cameleopardalis (Linnaeus, 1762)] and impala [Aepyceros melampus (Lichtenstein, 1812)] were highly susceptible to experimental infection with the virus. After an incubation period of 6 days the giraffe developed a hard, painful swelling at the site of subcutaneous inoculation. This swelling increased in size and became necrotic thirteen days after its appearance. A generalized eruption of small nodules also appeared in the skin of the inner thigs and in the mucous membrane of the lips erosions and ulcerations were also evident. The animal died on the 15th day following the first appearance of lesions.

The experimentally infected impala developed a local reaction as well as a generalized eruption of skin nodules and mouth lesions after an incubation period of 25 days and died on the 6th day of the disease. Two buffalo calves

[*Syncerus caffer* (SPARRMAN, 1779)], less than three weeks of age, failed to develop clinical symptoms after experimental infection. In view of the small number of animals available and the possibility of a passive immunity at such a young age, the results of this experiment can not be regarded as conclusive evidence of the non-susceptibility of the buffalo to lumpy skin disease.

G. Immunity to Lumpy Skin Disease

A natural resistance to infection, which is not associated with immunity, has previously been described as occurring in approximately 50% or more of all cattle exposed to natural or experimental infection. This resistance is responsible for the variable morbidity rate encountered in outbreaks of the disease.

Animals which have recovered from apparent or inapparent natural infection develop antibodies in their sera capable of neutralizing up to 3 logs of virus and are also resistant to reinfection. WEISS (1963) has shown that neutralizing antibodies will persist for at least 5 years and expressed the opinion that the immunity is probably "life-long". On the other hand, ADELAAR and NEITZ (1948) found that immunity was of short duration and that cattle developed a local reaction when their immunity was challenged 11 months after recovery. HUSTON (1947) also reported recurrent attacks of lumpy skin disease in the same animals. However, the validity of these observations is questionable, since infection by the unrelated "Allerton" virus, which causes a clinical syndrome indistinguishable from lumpy skin disease, was not known to exist at the time (WEISS, 1963).

Serial passage of the "Neethling" strain of lumpy skin disease virus in the chorio-allantoic membranes of embryonated eggs resulted in attenuation of the virus for cattle. At the 20th passage level, the virus failed to cause a generalized skin eruption or other signs of illness and only produced a local swelling at the site of inoculation in about 50% of animals injected. The local reactions disappeared within 4 to 6 weeks without any evidence of necrosis. Virus could not be reisolated from the blood or the local lesions of inoculated cattle, and there was no evidence of reversion of virulence of the attenuated virus by re-passage in lamb kidney tissue cultures (WEISS, 1960).

Following immunization of cattle with the modified live virus vaccine, circulating antibodies were detectable on the 10th day and reached a high titre by the 30th day, especially in those animals which showed local reactions. The antibodies persisted for more than 3 years. In animals which failed to develop local reactions, a rise in antibodies was difficult to demonstrate. These cattle as well as those with antibodies were resistant to challenge (WEISS, 1960).

A passive immunity derived from the colostrum and which persisted for up to 6 months has been demonstrated in calves from immune cows (VAN DER WESTHUIZEN, 1964).

In Kenya, a strain of sheep pox virus is used as a vaccine to protect cattle against lumpy skin disease (CAPSTICK and COACKLEY, 1961; MACOWEN, 1962).

H. Clinical Features, Pathology and Diagnosis of Lumpy Skin Disease

The incubation period following artificial infection was found to average 7 days (HENNING, 1956). After natural exposure, cases usually occurred within

2 to 5 weeks (HAIG, 1957). The symptoms of the natural as well as the experimentally produced disease have been described in detail by MacDONALD (1931), THOMAS and MARÉ (1945), VON BACKSTRÖM (1945), HENNING (1956), HAIG (1957), ALEXANDER et al. (1957), BURDIN and PRYDIE (1959), ALEXANDER and WEISS (1959), CAPSTICK (1959) and WEISS (1963).

In animals which develop overt symptoms, the disease is characterized by a febrile reaction, which may last for 4 to 14 days, accompanied by inappetence, salivation, lachrymation and a nasal discharge, which may become mucopurulent.

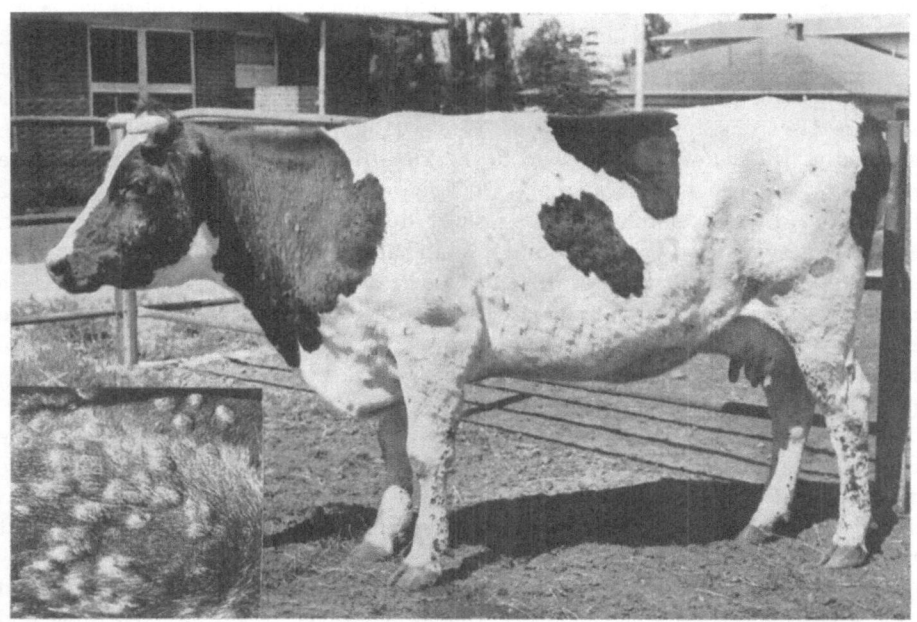

Fig. 8. Clinical appearance of skin lesions in an animal with lumpy skin disease. Inset shows a close-up view of the lesions. (Reprinted with kind permission of the Chief: Veterinary Research Institute, Onderstepoort.)

A generalized eruption of circumscribed, firm, round and raised nodules occurs in the skin of the entire body, particularly the skin of the neck, brisket, back, thighs, legs, perineum, udder and scrotum and around the eyes and on the muzzle, within 48 hours of the first rise in temperature. The nodules may vary from only a few in mild cases to several hundreds in severe cases and they are usually fairly uniform in size in the same animal, varying from 0.5 to 5 centimeters in diameter. Soft, yellowish-grey nodules may also appear in the mucous membranes of the mouth, nose, vulva and prepuce, where they macerate easily leaving erosions or ulcers. Lesions in the mouth are accompanied by excessive salivation and those in the nasal passages cause partial occlusion and laboured breathing. In some animals a conjunctivitis and keratitis may develop. The skin lesions are illustrated in Fig. 8.

Although indurated skin lesions may persist for many years, the nodules usually undergo complete necrosis. They begin to separate from the surrounding healthy tissue after about 7 to 10 days and become hard and dry to form sitfasts, which are eventually shed 3 to 5 weeks after the first appearance of the lesions.

The skin nodules are deep-seated and usually involve all the layers of the skin and even the adjacent subcutaneous tissues. Separation and early shedding of the necrotic portions of skin result in granulating craters of varying depth and size, which often suppurate due to secondary bacterial infection. Healing takes place by scar formation.

Apart from the circumscribed skin nodules, extensive subcutaneous oedema of the limbs, dewlap, udder, scrotum and vulva may be seen in a number of animals. These swellings may persist for weeks and the overlying skin, especially of the limbs, frequently becomes necrotic and sloughs off, leaving large suppurating wounds. Metastatic abscesses in the regional lymph glands, lungs and other organs may also occur.

A constant feature of the disease is the marked enlargement of the superficial lymph glands especially the prescapular, precrural and subparotid.

In lactating cows milk production ceases and if lesions are present in the glandular tissue or lactiferous ducts, a mastitis with permanent loss of secretory activity may result. Although bulls are usually temporarily infertile, a permanent sterility frequently results from involvement of the testicles. Affected animals lose condition and in pregnant cows abortions may occur.

It has been reported that lesions in the trachea may be followed by cicatrization, resulting in respiratory distress in animals many months after recovery (DE BOOM, 1948).

The mortality rate in affected animals varies from less than one to ten per cent (HENNING, 1956; HAIG, 1957). A mortality rate of 2 per cent has been reported in Kenya (BURDIN and PRYDIE, 1959).

Although the mortality rate is low, the disease is of major economic importance through indirect losses resulting from emaciation, temporary or permanent cessation of milk production, infertility in bulls and permanent damage to the hides (HENNING, 1956). The tanned hides of cattle slaughtered during the acute phase of the disease revealed blemishes, which adversely affected the quality of the leather (GREEN, 1959). The latter author drew attention to the striking resemblance of these lesions in the hides of cattle to those produced by the pox viruses in sheep and goats.

The *post mortem* appearance and distribution of lesions have been described by THOMAS and MARÉ (1945), HENNING (1956) and BURDIN and PRYDIE (1959). On section, the skin nodules appear as firm masses of whitish-gray tissue involving the entire skin and adjacent subcutaneous tissue, which is usually infiltrated with a reddish serous fluid. Where extensive oedema occurs especially of the limbs, the subcutaneous tissue is infiltrated with a yellow, jelly-like fluid. Circumscribed grayish nodules may also been found in the skeletal muscles, lungs, rumen and uterine wall. In the mucous membranes of the lips, tongue and other parts of the mouth, as well as the nasal passages, pharynx, larynx, trachea and bronchi, the lesions appear as a mass of soft grayish-yellow necrotic tissue which gives rise to erosions and ulcers. In the mucous membrane of the abomasum small ulcers may also be found.

A generalized lymphadenitis is always present and the malpighian corpuscles of the spleen, which may or may not be enlarged, are prominent. Petechial haemorrhages are frequently present in the omentum and capsule of the spleen. Petechial haemorrhages and degenerative changes in the liver and kidneys may occur (VON BACKSTRÖM, 1945; BURDIN and PRYDIE, 1959).

The histopathology of skin lesions has been studied by THOMAS and MARÉ (1945), DE BOOM (1948) and BURDIN (1959). In early lesions there is an acute focal inflammatory reaction and oedema of the corium and dermal papillae, accompanied by a perivascular accumulation of lymphocytes, histiocytes, plasma cells and some polymorphonuclear leucocytes and a local proliferation of fibroblasts. The smaller blood vessels show thrombosis, which is presumably responsible for the oedema and subsequent necrosis of the tissue.

Old lesions, which have progressed to the sitfast stage, show coagulative necrosis of the superficial epidermis and appears as a hyaline structureless material arranged in layers. Beneath this necrotic mass, the perivascular accumulations of mononuclear cells, proliferation of fibroblasts and thrombosis of bloodvessels can still be seen.

The presence of intracytoplasmic inclusion bodies in the epithelial cells, smooth muscle cells, infiltrating macrophages and lymphocytes and, occasionally, in the proliferating fibroblasts are of diagnostic significance. These inclusion bodies are eosinophilic and surrounded by a halo. They are variable in size, round or oval and may have globular excrescences. Some affected cells also show nuclear changes such as margination of the chromatin and juxtaposition of the nucleoli to the nuclear membrane.

The histopathological changes are illustrated in Figs. 9 and 10.

Although the clinical and autopsy findings of lumpy skin disease are fairly characteristic, it has to be differentiated from diseases such as onchocerciasis, besnoitiosis, demodicosis, streptothricosis as well as urticarial swellings, insect bites and bacterial phlegmosis. In view of the remarkable similarity of the early skin lesions of lumpy skin disease with those caused by "Allerton" virus infection, laboratory confirmation of the diagnosis is essential. This can be achieved by one or more of the following methods:

(i) Microscopic examination of stained sections of excised skin lesions preserved in 10% formalin and recognition of the characteristic histopathological changes and inclusion bodies associated with infection by lumpy skin disease virus (THOMAS and MARÉ, 1945; BURDIN, 1959).

(ii) Isolation of the virus in tissue culture. For this purpose, suspensions, prepared from excised nodules preserved in 50% buffered glycerine or from unpreserved scrapings of the superficial epidermis of the skin lesions, are seeded onto monolayer cultures of lamb or calf kidney cells. Although virus can also be isolated from saliva and blood collected during the early stages of the disease, unsatisfactory results are often obtained with these specimens (WEISS, 1963). In view of the fact that other agents, such as "Allerton" and orphan viruses, may be isolated at the same time and outgrow the virus of lumpy skin disease, stained coverslip or tissue culture preparations, prepared according to the method described by DE LANGE (1959), should be examined to identify the cytopathic effects of lumpy skin disease virus (ALEXANDER and WEISS, 1959). The cytopathic effects of Allerton virus resemble those of the herpesvirus group and

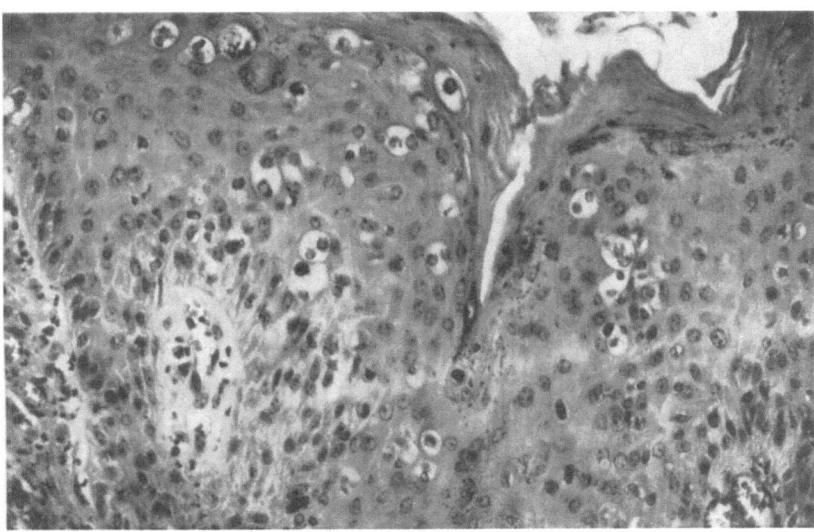

Fig. 9. Histopathology of skin lesion in lumpy skin disease. Staining: haematoxylin-eosin. Magnification 200 ×. Note cell infiltration and inclusion bodies in degenerating cells. (Printed with kind permission of the Chief: Veterinary Research Institute, Onderstepoort.

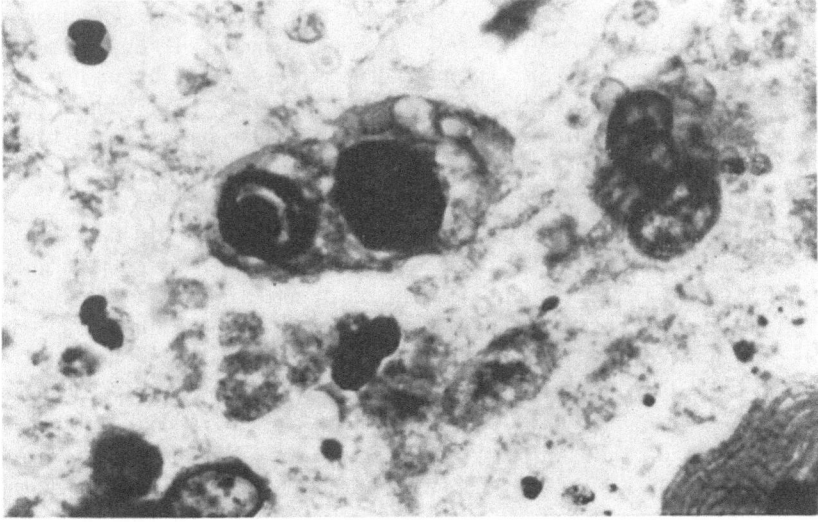

Fig. 10. Histopathology of skin lesions in lumpy skin disease. Staining: haematoxylin-eosin. High magnification: 1200 ×. Note large intracytoplasmic inclusion body in mononuclear cell.

have been described by ALEXANDER et al. (1957) and DE LANGE (1959). The isolation of orphan viruses from normal cattle as well as from animals suffering from lumpy skin disease, has been reported by ALEXANDER and HAIG (1956), ALEXANDER et al. (1957), PLOWRIGHT and FERRIS (1957), ALEXANDER and WEISS (1959), and PRYDIE and COACKLEY (1959). The cytopathic changes associated with these agents and which are essentially characterized by the formation of large intranuclear inclusion bodies, have been described by ALEXANDER et al. (1957), DE LANGE (1959) and PRYDIE and COACKLEY (1959).

The final identification of lumpy skin disease virus isolated in tissue culture is done by means of the serum-virus neutralization test described by WEISS (1963).

(iii) The disease can also be diagnosed in retrospect by comparing the antibody levels in acute and convalescent phase serum samples by means of the neutralization test. A rise in antibodies in the convalescent serum indicates recent infection.

J. Epizootiology of Lumpy Skin Disease

Although all the different modes of transmission of lumpy skin disease have not yet been established with certainty, circumstantial evidence suggests that biting insects may play a major role in dissemination of infection. The disease was found to be more prevalent during the wet summer months, particularly in low-lying areas and along water courses, where its spread could not be controlled effectively by quarantine measures (MacDONALD, 1931; MORRIS, 1931; VON BACKSTRÖM, 1945; DIESEL, 1949; HENNING, 1956; HAIG, 1957). Attempts to isolate virus from various species of insects, caught on infected animals, failed with *Culex* and *Aedes* species of mosquitoes, *Culicoides*, various species of ticks and *muscidae*. However, from *Stomoxys calcitrans* and *Biomyia fasciata* caught on infected animals, lumpy skin disease virus was recovered. By artificial feeding of virus suspensions to *Biomyia fasciata*, virus was recovered for at least 3 days after feeding. Although transmission experiments in cattle have yielded negative results, it is believed that insects, especially *Biomyia fasciata*, are capable of transmitting the virus mechanically (DU TOIT and WEISS, 1960).

Observations during the outbreak of the disease in Kenya also suggested that insects, particularly mosquitoes, were possibly involved in the transmission of the disease. Large numbers of mosquitoes, especially *Culex mirificus* and *Aëdes natronius*, were present on 55 of the 58 infected farms. In other respects, however, the evolvement of the epizootic in Kenya differed from that previously described in South Africa, in that the spread of the disease was slow and could be controlled by zoo-sanitary measures. In addition the morbidity rate was low and the clinical cases mild (MacOWEN, 1959; BURDIN and PRYDIE, 1959).

There is no doubt that lumpy skin disease can spread in the absence of insects and even under conditions where direct or indirect contact of infected with susceptible animals does not exist (HENNING, 1956; HAIG, 1957). The dissemination of the disease appears to be facilitated through movement of clinically sick or inapparently infected animals along the main roads or by rail (HAIG, 1957; MacOWEN, 1959). HAIG (1957) believed that the disease was contagious and was spread directly and indirectly by means of fomites. However, deliberate attempts to infect susceptible animals, through handling infected animals first, were un-successful (WEISS, 1959). These experimental animals were kept together in an insect-free stable, but were provided with separate feeding and drinking troughs. Successful transmission was only obtained when the infected and susceptible animals in the stable were watered at a common drinking trough, thus con-firming the suspicion that infective saliva might contribute towards the spread of the disease (HAIG, 1957). HENNING (1956) claimed that the disease was trans-missible to suckling calves through infected milk.

Birds and arthropods have also been incriminated as possible vectors of lumpy skin disease virus (HAIG, 1957; BURDIN and PRYDIE, 1959).

Following its first appearance in South Africa in 1944, lumpy skin disease spread throughout the country and became firmly established. Sporadic out-breaks of the disease have occurred throughout the ensuing years in widely separa-ted parts of the country, especially during the summer months, although fresh outbreaks have also been reported during mid-winter (HAIG, 1957). At irregular intervals of several years the disease has assumed epizootic proportions, probably due to the build-up of a sufficiently large susceptible population (WEISS, 1963), Although the disease also affected beef cattle in the ranching areas, most of the outbreaks occurred in the dairying districts with dense cattle populations.

Observations in Kenya, where sheep pox is enzootic, led to the suggestion that lumpy skin disease in cattle was introduced by sheep showing a pox-like disease (BURDIN and PRYDIE, 1959). By virtue of the experimental production of skin lesions in sheep by the intradermal inoculation of the Londiani strain of lumpy skin disease virus, and the development of skin lesions in cattle injected intradermally with the Isiolo strain of pox virus, isolated from sheep, CAPSTICK (1959) suggested that sheep might possibly act as carriers of lumpy skin disease virus. These observations have not been confirmed in South Africa. Sheep pox does not occur in this country and although there is close contact between sheep and cattle, there is no evidence to suggest that sheep are at all involved in the spread of lumpy skin disease. Not a single case of a pox-like disease has been en-countered among the more than 30 million sheep in South Africa since lumpy skin disease first appeared in 1944 (WEISS, 1963). It is therefore not known how the virus maintains itself in nature, except by cattle to cattle transmission.

During an outbreak of lumpy skin disease in a herd of cattle, some animals may escape infection entirely and remain susceptible (WEISS, 1959). The reason for this is not clear, but it appears that the disease is not highly infectious. This, as well as the natural resistance of certain animals and acquired passive immunity are factors which are probably responsible for some of the peculiar epizootiological features of the disease.

References

ADELAAR, T. F., and W. O. NEITZ: Unpublished work (1948). Cited by HAIG, D. A. (1957).

ALEXANDER, R. A., and D. A. HAIG: Unpublished observations (1956). Cited by HAIG, D. A. (1957).

ALEXANDER, R. A., W. PLOWRIGHT, and D. A. HAIG: Cytopathogenic agents associ-ated with lumpy skin disease of cattle. Bull. epizoot. Dis. Afr. 5, 489—492 (1957).

ALEXANDER, R. A., and K. E. WEISS: Unpublished observations (1959).

ANDREWES, C.: Viruses of Vertebrates. Baillière, Tindall and Cox, London, 1964.

BURDIN, M. L.: The use of histopathological examinations of skin material for the diagnosis of lumpy skin disease in Kenya. Bull. epizoot. Dis. Afr. 7, 27 (1959).

BURDIN, M. L., and J. PRYDIE: Observations on the first outbreak of lumpy skin disease in Kenya. Bull epizoot. Dis. Afr. 7, 21 (1959).

CAPSTICK, P. B.: Lumpy skin disease: experimental infection. Bull epizoot. Dis. Afr. 7, 51—62 (1959).

CAPSTICK, P. B., and W. COACKLEY: Protection of cattle against lumpy skin disease. I. Trials with a vaccine against Neethling type infection. Res. Vet. Sci. 2, 362 (1961).

CAPSTICK, P. B., J. PRYDIE, W. COACKLEY, and M. L. BURDIN: Protection of cattle against the "Neethling" type virus of lumpy skin disease. Vet. Rec. 71, 422 (1959).

DE BOOM, H. P. A.: Knopvelsiekte. S. Afr. Sci. Bull. 1, 44 (1948).

DE LANGE, M.: The histopathology of the cytopathogenic changes produced in mono-layer epithelial cultures by viruses associated with lumpy skin disease. Onder-stepoort J. vet. Res. 28, 245 (1959).

DE SOUSA DIAS, A., et J. LIMPO SERRA: La dermatose nodulaire au Mozambique. Bull. Off. Int. Epiz. 46, 612 (1956).

DIESEL, A. M.: The epizootiology of lumpy skin disease in South Africa. Proc. 14th International Veterinary Congress, London 2, 492—500 (1949).

DU TOIT, R. M., and K. E. WEISS: Unpublished observations (1960).

EASTERBROOK, K. B., and C. I. DAVERN: The effect of 5-bromodeoxyuridine on the multiplication of vaccinia virus. Virology 19, 509—520 (1963).

GREEN, H. F.: Lumpy skin disease: its effect on hides and leather and a comparison in this respect with some other skin diseases. Bull. epizoot. Dis. Afr. 7, 63 (1959).

HAIG, D. A.: Lumpy skin disease. Bull. epizoot. Dis. Afr. 5, 421—430 (1957).

HENNING, M. W.: Animal Diseases in South Africa, 3rd edition, C.N.A., Cape Town (1956).

HUSTON, P. D.: Southern Rhodesia. Report of the Chief Veterinary Surgeon (1945, 1947).

LE ROUX, P. L.: Notes on the probable cause, prevention and treatments of pseudo-urticaria and associated septic conditions in cattle. Northern Rhodesia Depart-ment of Animal Health, Newsletter, pp. 1—4 (1945).

MACDONALD, R. A. S.: Pseudo-urticaria of cattle. Northern Rhodesia Department of Animal Health, Annual Report 1930, pp. 20—21 (1931).

MACOWEN, K. D. S.: Observations on the epizootiology of lumpy skin disease during the first year of its occurrence in Kenya. Bull. epizoot. Dis. Afr. 7, 7—20 (1959).

MACOWEN, K. D. S.: Personal communication (1962).

MADIN, S. H., and N. B. DERBY: Established kidney cell lines of normal adult bovine and ovine origin. Proc. Soc. exp. Biol. (N.Y.) 98, 5744 (1959).

MARTIN, W. B., B. MARTIN, D. HAY, and I. M. LAUDER: Bovine ulcerative mammil-litis caused by a herpesvirus. Vet. Rec. 78, 494—497 (1966).

MORRIS, J. P. A.: Pseudo-urticaria. Northern Rhodesia Department of Animal Health, Annual Report 1930, p. 12 (1931).

MÜLLER, G., and D. PETERS (1963): Cited by MUNZ, E. K., and N. C. OWEN (1966).

MUNZ, E. K.: Personal communications (1965, 1967).

MUNZ, E. K., and N. C. OWEN: Electron microscopic studies on lumpy skin disease virus type "Neethling". Onderstepoort J. vet. Res. 33 (1), 3—8 (1966).

NAGINGTON, J., and R. W. HORNE (1962): Cited by MUNZ, E. K., and N. C. OWEN (1966).

PLOWRIGHT, W., and R. D. FERRIS: Cytopathogenicity of rinderpest virus in tissue culture. Nature (Lond.) 179, 316 (1957).

PLOWRIGHT, W., and R. D. FERRIS: Ether sensitivity of some mammalian pox viruses. Virology 7, 357 (1959).

PLOWRIGHT, W., and M. A. WITCOMB: The growth in tissue cultures of a virus derived from lumpy skin disease of cattle. J. Path. Bact. 78, 397 (1959).

PRYDIE, J., and W. COACKLEY: Lumpy skin disease: tissue culture studies. Bull. epizoot. Dis. Afr. 7, 37—49 (1959).

THOMAS, A. D., and C. V. E. MARÉ: Knopvelsiekte. J. S. Afr. vet. med. Ass. **16,** 36—43 (1945).

THOMAS, A. D., E. M. ROBINSON, and R. A. ALEXANDER: Lumpy skin disease: Knopvelsiekte. Onderstepoort, Division of Veterinary Services, Veterinary Newsletter No. 10 (1945).

VAN DEN ENDE, M., R. A. ALEXANDER, P. DON, and A. KIPPS: Isolation in chick embryos of a filterable agent possibly related etiologically to lumpy skin disease of cattle. Nature (Lond.) **161,** 526 (1948).

VAN DER WESTHUIZEN, B.: Unpublished observations (1964).

VAN ROOYEN, P. J., N. A. L. KÜMM, K. E. WEISS, and R. A. ALEXANDER: A preliminary note on the adaptation of a strain of lumpy skin disease virus to propagation in embryonated eggs. Bull. epizoot. Dis. Afr. **7,** 79 (1959).

VAN ROOYEN, P. J., and E. K. MUNZ: The optimal conditions for the multiplication of "Neethling" type lumpy skin disease virus in embryonated eggs (1965). To be published, Onderstepoort J. vet. Res. (1967—1968).

VAN ROOYEN, P. J., and K. E. WEISS: Unpublished observations (1966).

VON BACKSTRÖM, U.: Ngamiland cattle disease: preliminary report on a new disease, the aetiological agent being probably of an infectious nature. J. S. Afr. vet. med. Ass. **16,** 29—35 (1945).

WEISS, K. E.: Unpublished observations (1959, 1960, 1962, 1964, 1966, 1967).

WEISS, K. E.: Lumpy skin disease. Emerging diseases of Animals, FAO Agricultural Studies, No. 61, pp. 179—201. F.A.O. Rome, 1963.

WEISS, K. E., and J. BROEKMAN: Unpublished work (1965).

WEISS, K. E., and S. M. GEYER: The effect of lactalbumin hydrolysate on the cytopathogenesis of lumpy skin disease virus in tissue culture. Bull. epizoot. Dis. Afr. **7,** 243 (1959).

WEISS, K. E., and H. WEYLAND: Unpublished observations (1960).

WESTWOOD, J. C. N., W. J. HARRIS, H. T. ZWARTOUW, D. H. J. TITMUS, and G. APPLEYARD: Studies on the structure of vaccinia virus. J. gen. Microbiol. **84,** 67 (1664).

YOUNG, E., P. A. BASSON, and K. E. WEISS: Experimental infection of the Giraffe [*Giraffa cameleopardis* (Linnaeus, 1762)], Impala [*Aepyceros melampus* (Lichtenstein, 1812)] and the Cape Buffalo [*Syncerus caffer* (Sparrman, 1779)] with lumpy skin disease virus (1966). To be published, Onderstepoort J. vet. Res. (1967—1968).